FLY FISHING PRESSURED WATERS

-NTR-
TFO

FLY FISHING PRESSURED WATERS

Tactics, Techniques, and Strategies for Popular Western Rivers and Beyond

Nicholas Conklin and Capt. Collins D. Illich

STACKPOLE
BOOKS
Essex, Connecticut

To the many teachers who have impacted our lives, without whom neither of us would have been able to experience such a fulfilling, literature-filled life

STACKPOLE BOOKS
An imprint of The Globe Pequot Publishing Group, Inc.
64 South Main St.
Essex, CT 06426
www.GlobePequot.com

British Library Cataloguing in Publication Information available

Library of Congress Cataloging-in-Publication Data available

ISBN 9780811775670 (cloth)
ISBN 9780811777773 (electronic)

Printed in India

CONTENTS

Public land extends hunting, fishing, and hiking access to all. It is critical to our sport and the community that everyone is encouraged to enjoy these areas, respectfully. We hope to discover a way to allow more lands to be publicly accessible, so anglers can work together to protect the beauty of the natural world we have inherited. PHOTO BY JOHN MITCHELL

FOREWORD

I often find myself skimming through new books, sifting the gravel to find the garnets. Once in a while, however, I discover a book that demands a thorough read, even returning to reread. This is such a book.

In writing *Fly Fishing Pressured Waters*, Nicholas Conklin and Collins D. Illich have created a resource that is part wake-up call and part call to action, one that not only makes us better anglers in the current climate of our sport but also makes us more aware of the need to be better stewards of the treasures that have been entrusted to us for a short time. Through their writing, we learn to be more successful and responsible anglers. This book benefits readers individually by making us better prepared and efficient on the water and off, but it also benefits us collectively as we navigate a future in our sport that will surely put more pressure on our resources.

The L.A.S.T. approach to casting, with the emphasis on understanding the path of the rod tip, is a unique and effective learning tool. It should send us all out to the backyard to observe our rod tip path. The authors emphasize the need to spend time selecting the best rod for the job, whether one we own or one we're considering purchasing. They then underscore the need to practice real-life scenarios.

Line management, especially mending, is an oft-overlooked art. When done correctly, it enhances a drag-free drift. When done incorrectly, it yanks the fly out of the current lane, rips the surface of the water, and likely spooks the fish. I found their tips regarding line management and mending quite useful.

The thoughtful presentation on gear (including vest, packs, and fly boxes) provides valuable insight for choosing the best option for your preferences, even to the point of getting the right net. I found a lot of gems in the packing, repacking, and gear cleaning advice. I'm now hoarding moist towel packets and even repurposing desiccant pouches to throw in my streamer box and to pull moisture out of wet gear. I remember, a few decades ago, opening an old, disheveled box of fluffy, furry flies with rusted hooks and eyes that I had to throw away. I won't have to do that again!

In the course of this book, Conklin and Illich offer the best description of the complete angler since *The Compleat Angler*, whether drifting or wading, emphasizing the need for a game plan and preparedness. They describe an angler who is conscientious, observant, and aware of other anglers, ready for any contingency on the water.

This book is a must-read for all anglers, in part for our safety and enjoyment, but also for the safety and enjoyment of our fishing buddies. It will likewise help us fulfill our obligation to further the sport we love and to protect its resources.

With their years on the water as guides and instructors, combined with their many years in the fly-fishing industry, Conklin and Illich are uniquely qualified to share this information. Theirs will be an important voice in our sport for years to come.

—Jason Randall, author of *Nymph Masters: Fly Fishing Secrets from Expert Anglers*

ACKNOWLEDGMENTS

Writing this book with my best friend has been nothing short of incredible. Fishing some of my favorite waters and learning some new ones with an angler like Nicholas Conklin is special, and I am lucky to have had this experience.

Nick aside, this would not have been possible without the encouragement from my family. My wife Katie; our kiddos Walter, Susan, and Dobie; my mom Nanette; my brother Niles and sister Karli; and my dog Sony all pushed us and supported our late nights editing, writing, and planning ridiculous road trips all over the West.

Both Nick and I are lucky to have fished, worked, and designed products with some of the finest anglers in the world. In particular, I am fortunate to have watched and learned from Flip Pallot and Ed Jaworowski. Though they have vastly different casting styles, they are both masters of their craft.

Lastly, this would never have happened without my dad, Harold Illich. He introduced me to fly fishing after his time at the University of Montana. He encouraged and supported my fishing. Without Harold, I would never have picked up a fly rod. Go Griz!

—Capt. Collins D. Illich

This book, like many books, was an investment in time and endeavor of fun. We set out with a clear goal: to share what we have learned from others, from ourselves, and, often, from plain luck. What we wanted to create was a book not meant for the accomplished, tenured angler. While we hope they can mine the text for gems as they would dig around for a size 22 midge from an old, trusted fly box, we wanted to present a modern text about this sport and how the focus should always be on fun. While we feel we have interwoven these chapters with a fair amount of humor, we did want to present as many facts and as much knowledge as possible. We have always felt that too many books have a twinge of superiority. We did not like that. We aimed to approach it slightly differently; we hope we achieved that goal.

From sitting on a one-track bridge overlooking the Madison River to countless philosophical strategy sessions at the local bar, we have spent many hours crafting what we feel is something the industry needs—if not needs, just another way to consider approaching the sport, and the water.

I have many great people to thank, without whom I would have never emerged from my safe, wood- and water-saturated life to make the permanent move to Texas. While my acorn did not fall too far from the tree of skilled outdoors people, our paths seldom crossed and many of them passed on before I made the fatal mistake of getting into this dirty, fish-scaled business. I do owe personal gratitude to my siblings, parents, and all the folks I have met who have been as close as family and have contributed much to my life. I am likewise indebted to my partner Camille for her encouragement and support for the many hours and days spent in pursuit of this project and of an angling lifestyle.

No one book, video, or fly-fishing presentation of theories can ever be 100 percent accurate or the end-all, be-all in terms of information. Many terrific books have come before this one, and we passionately believe that if we are not all faced with economic collapse and subjected to a Mad Max lifestyle of basic survival, there will be more to come. The authors and anglers who have contributed to the vast libraries of fly-fishing literature deserve recognition. We have studied much of what they have written and how they have approached the sharing of fly-fishing information. What we have compiled about fishing pressured water is not new or meant as a rule, but rather thoughts based on experiences and some principles to help guide the way. It is our thoughts and experiences penned to help mitigate the barrier to success for anglers of all skill and experience levels.

I say "our" pointedly—anyone who works in fishing, and fly fishing specifically, has a responsibility to share knowledge, not show it. My coauthor and I wish only to share a bit about what we have learned from others and from many cold, frustrating, fishless days. With any hope, one of these sentences may compel even a self-proclaimed expert angler to try something new. Or, at the very least, take one of these thoughts and attack it with vigor, hoping to disprove the authors. We welcome that too.

I have been lucky for much of my professional career to work alongside fly-fishing greats, who have helped shape my many years spent on the business side of the sport. I would never have made it to this stage without the trust and kindness of those at Temple Fork Outfitters. While many kind, generous, and giving people have come and gone, the current ownership and management have shown nothing but support for my growth in the industry and this project. For that I am thankful. Many thanks to David Sams and the family at *Lone Star Outdoor News*, who took a chance on a young writer and helped me start my Texas journey.

Without my many years spent on the road, working dusty, dirty convention centers and drowning away the day's triumphs and tragedies, I would never have met such important mentors and friends. While the logo they wear on their shirt may change, I will be forever grateful to people like Jason Randall, Nick Curcione, and Ed Engle. All of them have shared a side of themselves and their gifts without asking. I truly consider them close friends and mentors despite age or distance. Without their mentorship and friendship, I would never have gotten even the initial lines of a book pitch scribbled out.

Many others contributed to this effort, whether in words, free ideas, questions, solutions to common angling problems, or time on the water. Our deep appreciation goes to our friend and incredibly talented photographer Oliver Sutro. We would also like to thank the crew of anglers and friends who volunteered their time and likenesses here. Our time spent capturing and crafting images was enjoyable and special due to their willingness to share their skills and time.

We would like to acknowledge several folks not necessarily tied into this insane circus called fly fishing. We appreciate the countless hours the staff at the Avenue in Dallas, Texas, allowed us to sit in their establishment while crafting the finer points of this text. They were always accommodating, and we appreciate that kindness.

—Nicholas Conklin

Those who lived close enough to fishable lakes and rivers were able to spend many uncrowded hours on the water during the lockdown. It was initially a solitary time, with little influence from other anglers; fish could evade harassment, and the resource flourished. As factual information about the causes and effects of the COVID pandemic came to light, anglers started to travel again, and fishing pressure increased. PHOTO BY CAPT. COLLINS D. ILLICH

INTRODUCTION

We, the authors of this book, met twelve years ago working together inside the sportfishing industry. Over that time, we have driven hundreds of thousands of miles together fishing across the entire United States. We are fortunate. Our partners not only support our passion but also encourage it. Every few months we load up a truck and head somewhere to explore new or longtime favorite waters. Many times, these trips take us to places we instantly fall for, and success comes with relative ease (whatever the barometer for success is that day). Other trips are meant to avenge the curse of previous struggles and fishless days.

On a recent trip to the West, we were doing our part in supporting the local establishments. Whether at the bar, boat ramp, or fly shop, we were constantly barraged with chatter about the "number of anglers" on the water. We spent the rest of that trip talking about fishing, fishing pressure, and the future of our sport.

We came to one conclusion: Angling pressure is only going to increase. That is a pain for the angler. But in the end, it's good for the fishery, its establishments, and the shops and their guides. Angling pressure protects the resource. We know that idea can be hard to digest as a sport angler, but look at the 1980s Texas redfish scene. Acres of fish diminished to nothing until the sportfishing anglers' attention was applied. Now it is a vibrant, sustainable fishery.

Our conclusion on that road trip was that angling pressure is not going to shrink. Our best strategy is to think about how we can better approach our waters with more anglers fishing around us while enjoying the time and angling success. That is the impetus for this book.

Our intent was initially just a conversation about one location or another. We spent most of the time on that road trip discussing approaches and how we could achieve our goals with a seemingly endless number of like-minded anglers around us. During that trip, we decided to start writing it down.

The cold, hard facts are simple: More people are fishing, and unless you have some killer trust-fund access to miles of private water, you and your approach need to adjust, or you will not enjoy your time on the water. *Fly Fishing Pressured Waters* is not designed to be a guidebook but more of a philosophical approach. Take some time, be thoughtful, and enjoy your time on the water.

We hope our book helps you do that.

—Collins and Nicholas

Although traffic was often sporadic during the COVID lockdown, anglers returned in hordes, crowding boat ramps and public access sites throughout the West. While activity increased, many anglers failed to adjust their approach to counter the traffic. PHOTO BY CAPT. COLLINS D. ILLICH

CHAPTER 1

UNDER PRESSURE

Fly anglers who have spent time around the boat ramps, public access sites, or dusty logging roads of the western United States have likely noticed a flurry of activity. During the initial stages of the COVID pandemic, most Americans faced hard realities or were at least troubled by the concept of an uncertain future. By the time most of us arrived at some semblance of comfort, there weren't any hard facts on how to return to our previous lives. We had to foster a "wait and see" attitude while doing the best we could to stay safe and healthy. Like the proverbial "travel safe" texts we receive before departing on a plane, we had no control, and we were along for one of life's most bizarre, back-alley taxicab detours.

We decided to exercise, dig our hands into the mud and garden, and tweak our overly optimistic checklist of activities we had always wanted to do (or at least endeavored to do since we learned of it that morning on Instagram). Could we safely accomplish these tasks within the confines of our homes? The outside world was a scary one, and we dared not venture beyond our property lines.

Those of us with local access opportunities found a new sense of calm and easy, unpressured fishing. It was initially a solitary time, with little influence from other anglers. Fish were able to evade harassment, and the resource flourished. Monitors of the collective fisheries hid safely from the headlines, and any soon-to-be catastrophic blunders were seemingly miles down the road. For a time, a pleasant (albeit uncertain) peace existed.

Those who had drifted from the sport found the time and ability to return to their former leisure-time activity. Fishing-related interests were reignited, perhaps not out of love, but rather out of a deep impending sense of doom (or sheer boredom).

We all felt the finger on the flashing red button of total societal collapse, and many of us were willing or unwillingly forced to reexamine our own lives. Former interests and passions once pushed to the back of the closet now seemed attractive again and worthy of a spot on the kitchen table. Squeezing in that extra hour or two in the office in the hopes of impressing a superior now seemed inconsequential. The value of late-night business meetings or happy hours seemed far less critical. Precedence was given to an early morning fishing trip or to a few hours spent reading up on the latest bug hatch, trout species, or new fly reel.

As anglers pulled rods, reels, and tackle from the back of the closet, something incredible happened. Those "keep busy to stay sane" activities were forgotten, and true passion activities reemerged.

For those already active in the sport, a few things broke from their husks and fluttered to the surface, beyond merely an increase in their fishing time. A large amount of time is spent piddling around the internet, looking at reports, studying the weather, and digging into all the most recent product advancements. Anglers far removed from their favorite trout waters took to the internet either in hopes of escaping the monotony of a work-to-live routine or to spend money on tackle like it was going out of style. What broke was a sense of servitude, which led to a desire to spend more time actively engaged, not just thinking about it or telling friends about the yearly trip.

However, this intense period of page four-level internet research and ravenous purchasing soon gave way to the desire to get out and actually fish. As restrictions loosened, more anglers took to the water. Those who were already on the water started to feel the squeeze. With a deeper sense of knowledge, those brand new to the sport who finally had the time, money, and energy to participate did so.

Since the invention of sportfishing, the number of anglers has been a frequent trope. In recent times, crowded ramps, empty fly bins, and "non-locals" have become points of consternation. It has become an easy excuse for many that the fewer number of fish landed is in direct correlation to the number of people fishing. The all-too-common reminiscence of "how it used to be" has done more to decelerate the growth of the sport and hinder modern resource management.

Those who were not part of the local intelligentsia were silently looked down on and their non-native dollars were begrudgingly accepted. Fears of outsiders tainting these bucolic fish towns were discussed openly at bars and fly shop counters.

During the COVID lockdown, those who made their living on the water had the time and energy to focus on what was most important, time spent enjoying the sport with family and close friends. Thoughts of frustrating float days and difficult clients soon drifted away as minor liner notes of another place and another time. PHOTO BY CAPT. COLLINS D. ILLICH

Underdeveloped, myopic quips at the expense of paying customers bulged out like belly fat over an overworked belt and pair of Wranglers. If you had a flat brim, a pair of flip-flops, and an out-of-state rental car, you were public enemy number one.

While the jokes have aged as well as the bag of potato chips hanging behind the bar, we feel the tide is turning on many of those beliefs. Because of advancements in modern angling techniques and tools, intense environmental constraints, and a rapidly growing population, we have found ourselves in unexplored territory. And new territory requires, if not demands, new ways of thinking.

The collective inability of anglers to effectively discuss the protections needed for our resources, how to manage increased participation, and how to encourage a new generation is not a novelty. Increased angling pressure and less open water continue to be hot-button issues at trade shows, fly shops, and river-town bars. These unimaginative conversations and negative sniping have snowballed from a few wrinkled, sun-splotched curmudgeons sitting around a table to a digital earworm hindering participation and turning potential anglers and water advocates away from the sport.

Among anglers, it would appear as though solitude and success have become impossible to find. It's not harder to find—it's just that other like-minded folks have sought out the same thing, while a separate contingent has vehemently stood in opposition to the growth of the fly-fishing community.

As an industry, anyone indentured to the manufacturing, retail, guiding, or outfitting end of the business now engages in a bizarre game of tug-a-rope. Manufacturers, sales outlets, and guides want the business and are excited by the recent increase in anglers and fly-fishing interests. However, there is a fine line. No one quite knows how thick or where this line begins—it's a constantly moving target, which is not easy to articulate.

Yes, we want to make and sell products, but when is too much? How do we keep the balance? How do we develop the sport healthily and sustainably? How do we protect sensitive resources? These questions are posed whenever anglers gather and reflect on our sport from a micro or macro perspective. How

As restrictions loosened and anglers became more comfortable gathering in the outdoors, many returned to the group fishing they enjoyed before the lockdown. Fishing with one or two other anglers is a time-efficient way to accelerate the learning curve on new or frequently changing water. PHOTO BY NICHOLAS CONKLIN

they are answered will define what is passed on to the next generation.

From an angler's perspective, modern resources have made it easier to find water to fish in a world that is constantly feeling more claustrophobic. Travel information is instantaneous, those hidden "secret" spots are not so secret, and searchers are no longer required to spend countless days exploring on foot. YouTube videos, podcasts, internet searches, and guidebooks make finding "fresh" water easier than ever. However, how and when we approach them hasn't seemed to track on the same evolutionary path.

What is great for discovery and the individual angler frustrates the local or the angler who "grew up" fishing that water. Anglers with some perceived ownership of their water can get damned defensive (and sometimes criminally mischievous) when someone with out-of-state plates suddenly parks in "their lot."

Fly Fishing Pressured Waters focuses on a skill- and mental improvement–based approach to becoming a more effective angler on pressured western US waters. Focusing on the Northern and Southern Rockies region, this book covers technique-specific, success-oriented tools and skills along with an approach that will help anglers gain efficiency and have more enjoyable days on the water. Whether floating or wading, we discuss techniques, tips, and realistic training aids to ensure anglers achieve success when fishing in front of or behind other anglers, chasing pressured fish.

Beginning in the Northern Rockies, let's dive into just how many anglers have taken to the sport in recent years, with some of the most compelling numbers being data-driven by license sales. For simplicity's sake, we break down the Northern Rockies as the states of Montana, Wyoming, and Idaho. Our research into the Southern Rockies consists of Colorado, New Mexico, and Utah.

At the beginning of 2021, officials with Montana Fish, Wildlife & Parks estimated an increase in fishing license sales, with nearly 132,000 sold by early spring. Digging deeper, the department estimated a 12.5 percent increase in resident licenses and an almost 60 percent increase in nonresident licenses from 2020. The same staggering numbers can be found throughout the Northern Rockies, with the Wyoming department dropping subtle hints to nonresidents to "stay home" on certain busy weekends.

A 2016 article in the *Montana Standard* reported that one 40-mile stretch of the Yellowstone River, which flows over 692 miles in total, endured an estimated 65,000 anglers between the spring and winter of 2013–2014. Research was conducted on the popular stretch of water between Livingston and Reedpoint. The report shows that most fishing pressure occurred from the Emigrant Bridge downstream to the mouth of the Shields River, around 30 road miles.

Shifting to the Southern Rockies, one of the states that typically sees massive migrations of new residents and second-home anglers is Colorado. Renowned guide and outfitter Dustin Harcourt, of Harcourt Fly Fishing, has spent the past thirty years fishing in and around Glenwood Springs, Colorado. He has been a firsthand witness to the increase in angling traffic, both pre- and post-COVID. Harcourt estimated that nearly 35 percent of local anglers have second homes in the region. Often referred to as "Little Texas," he noted that many smaller towns in his area have seen incredible booms of new and non-resident anglers, thus leading to increased angling pressure. However, he stated that these transient anglers have done great things for the local economies. Because they are deeply passionate about the fishing, they have helped shape policy and awareness of the struggles that his waterways have endured. Harcourt estimated that nearly 80 percent of his clients are from out of state, from locations like Texas, California, and Missouri.

After polling several well-established guides in Aspen and the Roaring Fork Valley, it was estimated that around 70 to 90 percent of clients were nonresident. While not vastly different from before 2019, the consensus was that traffic has increased and the occasional or seasonal angler is now a more permanent fixture in the Colorado angling landscape.

Delving into recent license sales in the state, the uptick in anglers is evident. From 2019 to 2021, license sales in Colorado grew by over 100,000. According to Colorado Parks & Wildlife, this increase is a combination of resident and nonresident purchasers. However, these sales numbers are slightly skewed by the fact that these licenses include a combination of items such as search and rescue.

Even well-tenured anglers take the time to observe each other, as they scan the water for not only hatches but also the movement of other anglers. Anglers Jason Randall (near) and Ed Engle (far) have spent many seasons fishing Tricos. There likely hasn't been a session or hatch that has played out the same, and the only constant is change.
PHOTO BY NICHOLAS CONKLIN

In Idaho, over 376,000 licenses were purchased by anglers in 2020 through October of that year, 50,000 more than sold in 2019.

Similar numbers have come from the state of Utah, with the Division of Wildlife Resources taking a deeper dive into the increased participation, citing both fishing and hunting licenses on the uptick. They cite these numbers as combined, as many anglers and hunters have opted for a dual license. From March 2020 to February 2021, the division sold 222,124 combination licenses (which allow the holder to legally fish and hunt in the state of Utah), compared to 183,358 from March 2019 to February 2020. A total of 417,883 fishing licenses were sold during the 2020–2021 period, an increase of more than 100,000 licenses from the year before. While these numbers cannot be cleanly delineated into a resident/non-resident scenario, fishing pressure in Utah has increased, and fish likely haven't bothered to check IDs as more wading boots have trudged into their habitat. Most public data is from the time period during and directly after the pandemic, but these numbers do not seem to be fading as we approach the late 2020s.

With a drilled-down point of view focusing on specific stretches of a river in the Northern Rockies, state agencies have begun to offer more specific numbers on how many catchable fish are present. Most conservative estimates typically cite around 3,000 to 5,000 catchable fish per river mile on the Madison in southwest Montana, which for a 183-mile river is an impressive amount. Call it 750,000+ trout. With the right approach, there are plenty of chances to have an enjoyable day on the water.

As mentioned above, these fish have seen many a Thingamabobber, every variation of a nymph, dry, or streamer, and a lot of activity. Life as a fish in any type of water is a challenge.

Pick any popular western river, and it quickly becomes apparent how the intense amount of yearly pressure can impact a fishery. Note that most numbers cited here were from the mid- to late 2010s, with data still rolling in during the initial COVID era and the boom in angling interest. We will discuss some angling numbers in the coming chapters, but the number of anglers from March 2020 to September 2022 far surpassed anything that has been collected from the West in many years.

The rapid rise in angling participation now stands in stark contrast to the ever-changing environment. We will refrain from lamenting the oppressive heat, variable water flows, and rise in dumbassery that occurs when humans take control, which inevitably leads to poor decisions, volatile interactions, and negatively impacted waterways.

While we are neither intelligent enough nor delicate enough to discuss in scientific verbiage exactly how the environment has changed, we strongly contend that it has, and unfortunately not for the betterment of the fish. As monitors and stewards of the resource, we must be thoughtful. There is only so much water and only so many fish. Temperatures are rising, and water levels are in a constant state of fluctuation. While we as humans can hide away in our homes, with the shades drawn and our A/C units cranking out ungodly levels of cold, crispy air, fish cannot.

Next time ask yourself, "Is it too hot to hook a fish, fight it, and land it?" Just think about sitting in big-city traffic: It's 110 degrees, your A/C doesn't work, your windows are inconveniently stuck closed, and you can get only a garbled, staticky weight-loss infomercial on the radio. Brutal, right? Just consider that's how a western trout feels in July heat.

The same analogy can be applied to water levels. Consider what fish deal with when water levels rise and drop during the year. We equate it to sitting in your living room and every few minutes the ceiling drops a few inches. As the day wears on, the ceiling continues to creep closer, and you get nervous. Just then your kitchen closes off. Slowly your entire house compresses, and you're stuck in a small section of your living room and the ceiling is still dropping in on you.

Water conditions are important to analyze, but so are the pressures from above. There is a well-known bluff on the upper Madison where eagles are known to congregate. It's a slow, deep undercut into the rock. If you want a show, hook a fish there. It is a near guarantee your struggling fish will be plucked from the river, and not by your companion with the net. There is no catch-and-release here.

High summer temperatures and low water are constantly impacting western rivers. Anglers need to always consider the pressures exerted on fish. It's a tough balancing act to enjoy the sport when it fits into your schedule while ensuring healthy fish when the next round of PTO comes up on the calendar. PHOTO BY NICHOLAS CONKLIN

As inconvenient as it is, it is about protecting the resource. Anglers tend to have a greater sense of appreciation for the water they fish and care about. Be thankful for the pressure. Without angler interest, these resources would be further depleted. Instead of getting ornery when a pile of kids clomp out into your favorite run and froth the water with their fly lines, find a way to become more active in its protection. Ensure it will be around for future anglers. Because you were once that wide-eyed, smiling angler, just happy to put a line in the water and enjoy the day.

If you take anything away from this book, other than a deep appreciation of our great writing skills and incredible sense of humor, ideally it is that despite all the pressure and constantly evolving weather patterns, there are tangible and repeatable steps anglers can take to maximize fishing success.

By developing a strong and adjustable casting foundation, effective pre-trip preparation, and organization or analysis of the water objectives, anglers can drastically distance themselves from the crowd. *Fly Fishing Pressured Waters* affords the angler the ability to grow beyond tackle and casting to a more all-encompassing angling approach. Whether a wading or boat-oriented angler, learning to observe and dissect the movements of fellow anglers and boaters is key. Understanding how to develop correct practice habits, doing something until you can't get it wrong, and retaining that information for when you actually get to the water will help you rise above other anglers.

Studying the angler in front of as well as behind you can be vital to future success. Most floating anglers will take the path they have seen others navigate, without variance. Anglers watching from the bank will observe the fly placement of those in front of them and assume that is where most bug life exists.

This book discusses how by studying the wider playing field and angler movements, from the way their boats maneuver down to how anglers move

The authors spend ample time studying maps to further develop their on-the-water plan. Here they quickly review their pre-charted float plan for the day. While digital resources are always best for the most up-to-date data, hard-copy maps can be critical in service-less areas. PHOTO BY OLIVER SUTRO

"Secret" spots are not so secret, as access to informational resources has grown. Anglers are not always afforded an empty boat ramp and local wildlife. PHOTO BY CAPT. COLLINS D. ILLICH

their rod tips, examining everything from the placement of flies in a box to effective knot tying, they can greatly improve their success rates.

Although we have presumably emerged from our shelters and caverns of indoor comfort that were the early 2020s, we have entered vastly unfamiliar territory as anglers. Boat ramps, parking lots, and online shopping carts remain loaded. River flows will change as often as we (should) change our underwear. Product availability will rise and fall with the latest techniques and materials, and the pressures on our western fisheries will remain. How we handle that, and the skills we build to differentiate ourselves from other anglers, will be critical—not only for our own success but also for the health and viability of the resource.

The bottom line is to observe and adjust. Throughout the remaining chapters, the overwhelming emphasis is on observing how others approach the water, while also understanding that many anglers just "copy and paste" their approach, using the style and techniques of others. While a good place to start, following the lead of someone who hasn't put in the work to better understand your water and how to fish it differently won't bring success. Don't just assume the angler in front of you is competent and confident in their approach. Be critical, be thoughtful, and enjoy your time on the water.

A foggy, dusky morning calls for a casual walk along the river. No reason to rush or wade with haste. Part of this fly-fishing journey is enjoying your surroundings, and success can be found in a careful, thoughtful approach. PHOTO BY NICHOLAS CONKLIN

CHAPTER 2

TIME MANAGEMENT

Speed kills! Or something to that effect was directed at us all in our school days when we were warned of the horrors of drugs. We were cautioned about the result of engaging in dangerous, lurid behavior. We'd end up on the streets shoeless and the scourge of society. Well, at least that's how it seemed, coming from the habit-lined mouth of Sister Mary Kathryn.

While we are not comparing the seriousness of the worldwide opioid addiction to fishing, similarities to the addictive nature of the sport do exist. The most loose-fitting analogy we can make is how preparation, whether in school or for the future, can be correlated to how an angler prepares for success.

When anglers fail to formulate and commit to a plan, the experience will suffer. Moving too quickly through a run and not taking the time to tie proper knots are habits all anglers are guilty of. While the infrequent angler may not see these habits as negatively impacting their experience, they do. Negative time habits establish patterns that diminish abilities. That hurried pattern then builds on itself, and anglers put themselves at a disadvantage. Even well-skilled, well-practiced anglers will suffer when ample time is not committed to certain steps.

This will often lead to a lack of successful outings. The lack of success then diminishes many anglers' willingness to venture out again. Combined with the added pressure of increased traffic on the water, or the truncated time windows that traveling anglers are faced with, fly-fishing melancholy ensues.

For anglers able to frequent the waters of the West, the rise in angling traffic can also have an attitude-diminishing impact. One creeping sentiment we aim to dispel is the overriding negativity surrounding anglers crowding rivers and fishing

Although locals may come off as grizzled and unfriendly, fishing towns subsist on the angling crowd, and more times than not they enjoy the arrival of anglers. Never hurts to say hello and buy someone a beer.
PHOTO BY CAPT. COLLINS D. ILLICH

the places some folks inappropriately feel is "their water." This dark cloud of negative energy has dissuaded the curious angler from more popular waters in Montana, Colorado, and Wyoming. Regardless of the state and location, and despite what the old-timers may say, quality fish can still be caught.

Most western rivers have remarkably high fish counts, as many years of hard work and science-based efforts have effectively managed these habitats. Much has been done to help combat the onslaught of ever-changing environmental conditions. Fish towns have been caught in the catch-22 of this scenario ever since fly-fishing travel in the West began. While the tourism dollars benefit more than just anglers, residents resent the traffic in town, the number of people, and all that accompanies the seasonal influx of visitors.

For anecdotal purposes, we will highlight an extremely popular southwestern Montana trout stream. This river is iconic not only because of the

Whether hauling anchor or packing gear, anglers should always be cognizant of how much energy they need to expend. It can be a struggle to keep the adrenaline from impacting your fishing; a deep breath in these scenarios will improve your angling life. PHOTO BY OLIVER SUTRO

fish population, scenic views, and accessible water but also because of the number of anglers who fish it every year. At last count, the Department of Fish, Wildlife & Parks estimates between 2,500 and 5,000 fish per mile. If the total number of anglers visiting this river can be estimated in the tens of thousands each year, one could surmise just how many flies and presentations these fish see throughout a year. Numbers are most often manipulated to suit the objectives of the author. While trying to be as unbiased and impartial as possible, we endeavor to explain that there are a lot of fish that will take your fly. Despite the claustrophobic feeling you may get on seeing a crowded parking lot and a flotilla of fiberglass, with proper preparation, you can still be highly effective and successful and will have a fun time if you just GO fishing.

Once the mental and physical barriers of public opinion and the overriding sense of heavily crowded, highly educated fish are pushed aside, the real fun can begin.

The first step to achieving success in highly pressured western rivers is prioritization of time. By understanding how best to utilize time and energy, the infrequent angler can easily rise above. This elevation will be in both skill and overall enjoyment of the sport.

Conversely, a poor understanding of time, rushed knot-tying, and a general lack of paced preparation may lead to some lonely afternoons at the bar and power smoking in the parking lot. The development of steps and time-based procedures will lead to the required ossification of skills.

Preparation Speed

For most anglers, the rush and excitement of getting on the water is a feeling that helped instill an interest in the sport. Hurried movements and frantic thoughts before fishing are natural. While many who take to the sport reference the peace and serenity associated with wetting a line, those anglers would be at a loss to explain away the rapid heartbeat experienced when fantasizing about the next trip. But that excitement and anxiety can be the root cause of many hurried mistakes.

Deliberate movements put anglers in a better position for success, not the luck-driven occurrences that may have led to a few hooked fish in the past. In these pages, we offer a menu of skills and practice routines to help eliminate the element of luck and truly provide anglers with a path toward skill improvement. Whether a once-a-year western trout fishery visitor or a local, all anglers can benefit from refining how they prepare, how they behave on the water, and what adjustments they make after a trip.

It starts with preparing before your trip: how gear is arranged, what is packed or left behind, and the speed at which those steps are taken. In chapters 7 and 8, we will delve deeper into specific packing techniques and strategies for both the wading and the floating angler.

The focus here should be the pacing and patience in preparing for the day. By allowing more time to

When anglers focus and slow down, avoiding the right out of the car to the water approach, proper knots and rigging can be ensured. Here some anglers spend a bit of time checking knots and tying on flies before approaching the water they want to fish.
PHOTO BY OLIVER SUTRO

When discussing fish per mile and angling pressure, few places are as prolific as the Three Dollar Bridge section on Montana's Madison River. But with that fame comes crowds and educated fish. So how will you approach these educated, pressured fish differently from the angler in front of or behind you? PHOTO BY CAPT. COLLINS D. ILLICH

prepare, you will find that critical tackle will not be left behind, knots will be cinched tight, and mornings will begin much more comfortably.

Choosing Tackle

The authors of this book are fishing friends. We have been fortunate to spend the last twelve years in various western and international waters. Despite the amount of angling time accumulated over the years, we have vastly different approaches to how we manage our tackle. While there have been times when a last-minute trip has popped up and a pile of tackle has been hurriedly dumped into a bag, we both have learned the value of slow, thoughtful packing.

On a recent trip to Africa, Collins was faced with a baggage dilemma. With limited space and faced with learning a fishery requiring both fly and conventional tackle, serious thought had to go into what was carried. With an extended family trip shortly after a coastal fishing opportunity near Tanzania, and near nonexistent time between international flights, an excessive package seemed foolish. Two weeks before departure, flies and lures were aggressively collated into two large tackle trays. Rods and reels were thoughtfully culled, and everything was packed down into two small carry-on bags. While the temptation was to load up a Barnum & Bailey–style railcar full of tackle for every possible scenario, a calmer, more thoughtful head prevailed.

For personal western fishing, the quantity of tackle could not be more dissimilar. One author prefers to travel with as many well-curated rods, reels, and loaded fly boxes as would make the most ardent tackle collector envious. The other takes a more stick-and-bindle approach with a basic set of rods and reels and a single box of confidence flies, carrying this set of gear whether floating or wading. There are seldom discernable differences in success.

While divided in ideology, the linking factor is the time spent analyzing what is brought along and how that tackle is used. It should begin with the rod and reel, working down in a prioritized list.

Above: Rushing when assembling or disassembling tackle will only lead to tangles, twists, and knots. Here one of your authors was a little too anxious while hopping between fishing spots. He didn't take the time to store his rod and now has wasted significantly more time than necessary. Rod sleeves, hair ties, and neoprene wraps are simple solutions for keeping rods clean and untangled.
PHOTO BY OLIVER SUTRO

Right: Rod cases, bags, and all the other tackle that some anglers require for a week's trip. Hard cases with separators and labels can help maintain strict organization when heading to and from fishing the West. These also provide the safest way to travel with expensive tackle.
PHOTO BY NICHOLAS CONKLIN

Rigging Rods

Whether wading or floating, committed rod-rigging sessions before hitting the water will lead to more proficient time on the water.

Unfortunately, this lesson has been learned the hard way after years of fumbling. While our overall tackle allotment may exceed that of a Bass Pro Shop, how we prepare it has been scrutinized. The foundation is understanding how much more efficiently you can pre-rig before heading to the water, doing it the evening before. This relaxed, undistracted environment will allow for proper focus, which can be hard to achieve when standing riverside and fish are nipping at the surface. The distractions of rigging while standing in a parking lot will lead to misaligned rod sections and line sagging from a missed eyelet. Rushed knots and last-minute fly changes are all too common.

This is when the night before, or even days before, a trip can prove fruitful. Use this time as foreplay for the next day's fishing.

This time away from the river often prevents panic about missing a fish or time-sensitive bug hatch. Time taken to communicate and discuss rigging options and plans can be the foundation of a positive fishing day. Deciding on the next day's approach and what is to be fished is a great, calming way to organize personal tackle. Not only will this help make for a smooth morning, but it can also eliminate left-behind items or a missed fly guide when rigging a rod. Organizing your gear in a well-lit environment, with all the pieces in front of you, is far more efficient than struggling with early morning light or a headlamp.

Rigging rods the night before is a fantastic collaborative opportunity. Discussing the plan with fellow anglers and thoughtfully planning together can add

Preassembled and rigged rods allow the angler more time to watch the water and get other essentials in place. Whether floating or wading, having a set of pre-rigged rods with fresh leaders and clean fly lines allows for quick fly attachment when it is time to hit the water. PHOTO BY OLIVER SUTRO

Whether it's over beer and chicken wings or in front of a roaring fire, anglers, especially those in new or infrequent locations, always need to talk strategy. Waiting until the day of fishing to discuss goals and strategies leaves too much room for failure. PHOTO BY OLIVER SUTRO

to the overall experience, regardless of skill level. Even if all parties are committed to fishing the same tackle, the "why" can be verbalized.

It is best to avoid waiting until the trip to develop the approach. Foggy, potentially hungover minds or just a disparity in experience in the group can lead to a half-hearted plan and inevitably an angler or two following the more confident in the group. Understanding what has been done in the past, what others do, and how to approach it differently can be a deciding factor in success.

The Boat Ramp Dance

The morning bustle of activity at a western boat ramp can be as frantic as that of a major international airport. Not everyone operates at the airport at the same level of experienced professionalism. However, the adherence to precise departure times should always be manageable for the fly angler. That is to say, don't stumble out of the truck fumbling with your gear and rushing around like a frazzled parent, overwhelmed by the prospect of a three-hour flight to Disney World. You can actually control this experience as an angler.

Arriving prepared for the launch or pull-off, with tackle rigged and easily transportable, sets a powerful standard for the day. No running back to the vehicle for the forgotten item or "extra rod."

Panicky movements to and from the car can only cause a lengthening of the space between the angling companions' communication. Like a sadomasochistic nylon-, wool-, and neoprene-clad Tinder date, this "feeling out" stage distorts time that can be better spent establishing a plan of action. Without concise and direct communication, an angling team is sure to experience mediocrity.

The opening questions should be covered well before any float or wade trip is made. The assertive angler who wishes to rise above the fold should generate a hard list of wishes and goals before fishing. This list, while not always feasible, is a start to

In preparation for several days of fishing and floating, two well-prepared anglers spent time scouting a local boat ramp and discussing plans for the upcoming float. Although they wadered up and the boat was hitched, their goal was to dry run a few launch spots and gauge water depth, speed, and overall angling traffic. PHOTO BY CAPT. COLLINS D. ILLICH

making the most of the time on the water. If floating with new friends or a guide, these baseline conversations should be had well before getting a hundred yards downriver. The same occurs with the wading crowd after they lethargically step out of the car, confused about who is in charge. Skills and styles have not been discussed, so more time is wasted lingering around the lot than in the water. It can be equally disruptive for those who bound out of the car like a German wirehaired pointer after a bird. Anglers need to meter their enthusiasm of weeks of telling their friends they are "going fishing" and ensure that they are composed and focused before starting.

Who oversees the plan? Do we abide by the guide's plan? How do I communicate to the guide how I want to fish? Is it appropriate to offer boat-launching assistance? Offer to back the trailer? Or do I stand off to the side like a scolded child? The uncertainty of who does what and when serves only to delay success.

The boom in raft and boat sales in the early 2020s has been both a blessing and a curse to the western river. While it has done wonders for manufacturers, it has also given the keys to many unprepared rowers. While we dive into the floating angler in chapter 5, the competence and experience of an oarsperson can be another cause of the awkward shuffle during a launch.

Not only is it the cause of ramp backup, but when new water is staring someone down, congestion can also occur. This is when the team attitude and a coordinated plan can get the boat unhooked and out of the way of others. This cohesion can be achieved through practice. Professional guides are smooth and efficient because of practice. They didn't emerge from the womb with inherent trailer and boat management abilities. It was learned over time.

The patience and time spent managing a boat and understanding how to launch are key to developing a smooth program. The confidence to explain this program and direct traffic will only expedite the experience. Rowing and boat handling classes are common in major fishing areas and are the best way to build

Above: A smooth load and launch program both sets the pace and ends the day on a safe and smooth note. This can't be done without talking out a plan and each angler having a role, even if that role is standing to the side and letting the captain run the show.

PHOTO BY CAPT. COLLINS D. ILLICH

Right: Learning to launch takes time and experience. Don't wait until a busy Saturday in the middle of the summer rush to learn how to maneuver the boat on and off the trailer. Even practicing in the driveway or a parking lot with clearly painted white lines can help with the ease of trailered boat moving.

PHOTO BY CAPT. COLLINS D. ILLICH

skills. While we will touch on how to approach the water more safely, part of it will be developed by time spent on the water and time spent with experienced rowers. A chapter in a book unfortunately cannot convey all there is to proper handling. We will, however, attempt to highlight the importance of our practice and the clear extrapolation of a load and launch program.

Knot Speed

Quality and quantity of knots, which can be repeated, is a concept most anglers should embed into their DNA. Attempting to lash together a few knots in the parking lot or a few minutes into a float is not a recipe for success.

Once all the pre-tied knots have run their course or it's time to change flies, how well can you adapt? Excitement, nerves, and frustration from the morning drive can all impact preparation. Now that you are finally on the water, how well can a new fly be tied on, and how precise is it?

It will take an angler only one drift with a pigtail on a hookless line to know they have not securely finished the knot. This is not the fault of the leader; it is a failure of the angler.

When constructing a list of knots appropriate to the target, simple, repeatable knots will far outperform those with high advertised breakage rates. An encyclopedic recall of knots is not helpful either. If they cannot be tied in various (and often foul) conditions, they should not make the list.

While the lead-off knots were finished in the warm confines of a well-lit room, over a comforting drink, the real-time angler is far from that comfort. What further separates anglers is the speed and consistency with which they can retie a fly and return it to the water. Knot management is even more of a differentiating factor in western fishing, as most will be done with a multi-fly rig—double

Less knots, more confidence. The ability to tie knots and retie in poor conditions is key to quickly returning to active fishing. Unfortunately, too few anglers spend time perfecting a short list of knots that they are quick and confident with. PHOTO BY OLIVER SUTRO

Tying knots in the boat and other moving scenarios takes practice. No different from standing in the parking lot, anglers need to focus on two or three confidence knots. Even when anchored, this is not the time to cheat on a knot.
PHOTO BY CAPT. COLLINS D. ILLICH

nymph, hopper-dropper, double streamer, streamer-dropper—and most anglers will find a way to cheat on a knot or tangle the rig within five minutes of fishing.

What this step-skipping or lack of practice does is slow down the fishing. Time is wasted sitting down or standing mid-river fumbling with the line. Time is key, as is the ability to shut out outside forces and focus on the task at hand.

The wise angler should select at least three knots when building a system. First is a knot that can accommodate lines of varying diameter. This connection knot can be used when connecting a fly line to a butt section of material or a butt section to a thinner-diameter tippet.

The second style of knot should center on connecting the leader and tippet material—a slim profile knot that can be cinched together, whether fluorocarbon or monofilament. These materials will have a distinct level of slickness, so practice with both is wise.

The third knot should be a connection from the tippet material to the fly. Here the ability to tie a loop-style knot and a tighter cinching knot will cover many fly techniques. A loop-style knot will allow for freedom of movement, while a tighter cinching knot is ideal for both the eye of a hook and the shank. This is important for effective multi-fly approaches.

Boat anglers often have the advantage, as the spatial freedom and comfort to retie simply makes it easier. It also affords the ability to pre-tie several multi-fly rigs before heading downriver. This way, when the inevitable tangles occur, one quick knot to the tippet material is all that is needed. The best option is to use that fancy foam pad with the $29 price tag from the local fly shop to coordinate some replacement rigs. Pool noodles or those foam rollers used at yoga studios make great tools for pre-rigging droppers. Their roundness will prevent any kinks or nicks in the rig. They are easy to cut to preferred sizes, and the material allows for secured hook

While it may be a tough pill to swallow, effective anglers will step out of the run and retie flies after landing a fish or simply deciding to change the offering. Rushing or trying to detangle flies standing waist-deep in the water is only asking for trouble. PHOTO BY NICHOLAS CONKLIN

Anglers should be creative when searching out ways to make fly storage and general tackle storage easier. A simple trip down the aisle of a fly shop or a Target store can provide some clever options. Round, semisoft materials like a pool noodle or yoga roller act as great multi-fly rig holders. PHOTO BY OLIVER SUTRO

points. Rolls with ridges or cutouts will also help keep the tippet away from snags or possible contact points that could create a weak spot.

For wading anglers, you may turn back to the trusty plastic baggie. With the same approach of pretying two flies, you can place a handful in a baggie, spooled up like a tapered leader with a small zip tie. The zip tie will keep these untangled in the bag. This is also a way to repurpose the transparent plastic bags that tapered leaders are packaged in. Zippered tops allow for a more secure fit, and the clear plastic provides a quick view of what's inside.

Retying these rigs is not a matter of when, but rather how often. When determining when to "give up" on a fouled rig, it's best to not prolong the fight longer than 20 seconds. While you may not let the piece of cheese dropped on the kitchen floor sit that long before picking it up and eating it, anglers should not commit much more time to untangling a rig. They need to impose serious self-control when deciding to cut off and rerig versus working out the twists and tangles. Every few seconds spent untangling that mess is less time actively fishing. With a few pre-tied rigs, however, the daunting nature of retying can be eased.

Casting Approach

Once the above phases are consistent and there is a level of comfort in the wading approach, the next focus should be on the energy expended while casting.

Now is not the time to begin pushing too hard. While the temptation to cast harder may grow as a fishless day takes shape, the energy expense will not yield a linear outcome. Like patience and thoughtful wading, the casting effort should remain consistent throughout the course of the day.

Despite the initial excitement of wading into a run and hoping to bomb a long cast, effective anglers take their time, eventually lengthening out line as they fish the close water. Deliberate strokes and efficient use of energy will keep you on the water longer. PHOTO BY OLIVER SUTRO

In chapter 6, we will cover the casting approach and how best to prepare and correct mistakes on the water. This section will focus on the mental approach.

Casting with a river full of spectators is vastly different from a lesson in a parking lot or park from a casting instructor. But that casual, positive mental approach when trying out a rod should be no different when in a real-life fishing scenario. Forgetting to breathe and enjoy the day, unfortunately, fogs the experience for too many anglers. Everyone should remember that this is a leisure-time activity.

While we dig into the different casting and line management styles of various techniques, the efficient and effective angler needs to always consider the effort and attitude being put forth. If the effort and energy you are willing to contribute fluctuates during the day, perhaps a different technique should be examined. If certain rods require more work, then a deeper dive into the right rod action for your angling style is needed.

Orienting Tackle in the Boat

Serious thought is required when coordinating the layout of the boat or raft. There is limited time during the float to find a missing tippet spool or a box of flies. Rowers have the creeping concern of constant danger while trying to keep anglers properly rigged and in good casting lanes.

With all the concerns of just getting down the river, new boaters often don't take the necessary steps to properly load and organize the boat. A simple footprint, in which boat bags and rods can be easily pulled and replaced without much thought and movement, is key. Items should be located by priority, with the most touched items being as close to the dominant hand as possible.

A dry run is always recommended before the season begins. As new tackle and storage devices are accumulated in the off season, it is a wise practice. With a hearty imagination, rowers should mentally run through all scenarios before getting the oars wet.

Simple yet efficient placement of gear, in front of your line of sight, will ensure that everything you pack is ready when you need it. While hiding occasionally used items away can save immediate space, it's very easy to forget where certain items are hidden. PHOTO BY OLIVER SUTRO

Even late season fishing in the West will force anglers to encounter each other. Don't shy away from bright sunny days and cold weather spells if you are in search of less pressured water. PHOTO BY NICHOLAS CONKLIN

During especially long winters, we will even generate faux scenarios to test the efficacy of a new boat plan. Acting as a persnickety and bumbling angler, I (Nicholas) will post myself in the bow of the boat and run through several situations and requests that can occur. Throwing out as many challenges as I can think of, we time how long it takes to grab a specific fly, indicator, or other needed items. Additionally, I shout boulder left or right, to further develop a realistic sense of a float. With mimed rowing motions, not only can the rower keep the multitude of tasks in mind while navigating the river, but we can also refine our teamwork. This is the time to refresh our team-style approach and practice what situations we may run into come spring on the Madison River.

To keep a clear mental list of items, it is advised to never pull or remove tackle from the boat to utilize in other areas. While we always strive to be minimalist and have a streamlined approach, this is one scenario in which duplicate items should be stored and not leave the boat. No different from packing for a wade trip, this is one of the easiest ways a floating angler can misplace a critical piece of gear. In chapters 7 and 8, we cover how labeling tackle is one of the surest ways to effectively speed up organization. For the boat angler, this can be as simple as clearly labeling boxes and bags with a label maker.

As both an ardent wader and a boat angler, Collins even goes as far as labeling which bags and boxes belong to the boat or the hip pack. This is done when the season ends and the tackle is pulled from the boat to be cleaned out and stored. Although stored in the appropriate storage bins, a simple shorthand can ensure that when the snow melts and the tackle gets pulled back out, it will find its home. This labeling system is also a way to remain honest. It can be quite easy for boxes and spools of tippet to get shuffled around during a hectic season. With a simple step, this gear can be identified if it makes its way over the gunnels and into the back of the truck or pack.

Fishing When Others Won't

With all the traffic, busy waterways, and ancillary things that make effectively fly fishing the West more difficult, one of the most damning but easily correctable is the physical approach.

While a dry-fly purist may sit for hours on the bank, watching bugs and gauging the weather, they are still in waders and in the mix. The streamer angler may spend hours working a run, but they still have a fly in the water. Both are looking for their window and, most important, putting the time into fishing. Even if they are not catching anything, they are honing their skills.

It is important to realize that certain seasons and hours will be more productive compared to others. What is more important is that anglers who don't make the effort to fish in the preconceived "off hours" will not advance their success rates. While futzing around the house or hanging around a boat ramp is one way to while away the heat of the midday sun, the intrepid, adventurous angler will explore an outside-the-norm time window to learn for themself what opportunities exist. Ignorant of those scoffing at fishing the left bank, in high sun, the more flexible angler can have an open pick of the water that may be heavily fished in the mornings or evenings.

Either way, your time on the water and when you fish can have an important impact on your success.

Depending on the season, water flow, and the previous winter's snowpack, you may be faced with some daily restrictions. In positive-water years, and with no mandatory restrictions, anglers still tend to fish typical hours.

In early spring, the boat ramps are a ghost town until about 10:00 a.m. As the bright heat of a western summer attacks like a swarm of angry wasps, most anglers shift their water time to early mornings and late evenings. If the location and instillment of Hoot Owl and other fish protection measures don't keep anglers away, the rubber tube hatch will. Seasonal focus on off hours, such as early morning adventures before the first snows pack the high mountains or late-night sessions under a headlamp, can lead to less pressured fish. These trips can also garner much more information on the actual fish behavior of the

Cold, light-snow days like this one can pop up in early fall in the West. While it may be tempting to stay inside watching football, anglers can find quality water with bugs if they take the chance. By day's end, most of the snow had melted and the sun returned, activating fish feeding activity. PHOTO BY OLIVER SUTRO

While it may be slick and tough wading during the day, these challenges will only mount in low-light conditions. A slow pace and walking the stretch of water you plan to fish either late in the day or early the next morning will help build confidence and safe habits. PHOTO BY OLIVER SUTRO

targeted water. Don't be afraid to fish late into the evening and even into the early morning. Fishing late isn't just a responsible approach to the resource but also incredibly fun.

For out-of-town anglers, a creative approach to fishing on the clock will help keep you off the radar of those local mental defects, and soon-to-be-incarcerated fools, who don't care for a stimulated economy and the survival of local businesses. This has popped up in and around the West. People who slash tires based on license plate colors belong in the lowest circles of hell, next to the snotty, impotent purists and those who putter in the passing line.

On one late November Montana trip, we had stellar stream conditions and bug activity. Uncharacteristic of the date, temperatures were in the 70s, and we started out fishing in T-shirts. As happens in the West, by week's end we were blanketed in over 12 inches of snow. However, we approached the trip with the idea that most seasonal residents would be out of the state, and there would be a limited cadre of floating anglers.

Fishing late in the day also has benefits outside of just lighter traffic. Growing up in Michigan, I (Nicholas) spent many late summer nights fishing smaller, deeper, sandy-bottomed rivers. I learned the benefits and struggles of nighttime river fishing.

The same river I had fished a thousand times in the bright daylight, that I could wet-wade with my eyes closed, is never the same after the sun goes down and you have limited visibility. Yes, you may get lucky on a moonlit night, but more often than not you are operating blind.

I've picked up many lessons from my years fishing the Pere Marquette. For one, choose a section of the river and study the ins and outs of it. Not the water, but your path to the water. Get acquainted with the parking lot and whatever path you need to walk into the water. This takes time, but arriving early and walking the path, sans rods and tackle, will make your life easier.

Pay attention to the branches, tree limbs, or whatever obstructions that may slow you down when walking to a spot. Hiking down a trail near sunset, a

Small-creek fish on warmer days should command respect and deserve to be kept in a cool, wet net. This little brown afforded the angler a touchless release and smoothly slid from the net. This is when pinching the barbs on flies makes for a safer release for both parties. PHOTO BY NICHOLAS CONKLIN

fully rigged 9-foot rod will undoubtedly catch some sort of branch. You don't need to spend 10 minutes rigging up just to have to redo it by the time you get to the water.

Learn the water. Find a section and study it. Watch and track how the water flows at various times of the year. Keep the section small, starting with a portion around 10 to 30 yards wide. Are there long steady runs, big boulders, and/or drop-offs?

Note where and when you hook and land fish. They say, "You never fish the same river twice," but paying attention to fish location as you master your micro beat will translate across other fisheries.

Awareness of obstacles and understanding the distance between you, your fishing lies, and potentially disastrous obstacles is key. Your mastery and intimate understanding of a small section of water will make you a better angler with a deeper understanding of fish behavior and habitat.

Practice distances. If your water dictates fishing close during the day, you'll need to fish even tighter at night. Get comfortable covering the close-in water. This rule applies to most western rivers. When the sun goes down and the threat of predation decreases, the close-in lies and boulders will become a friendlier feeding lane.

For safety and comfort purposes, wading a little bit shallower will ideally keep you closer to the bank, and safer. Western rivers are deceptive. While the water may be only above the boot tops, it is still powerful and can be dangerous.

An old trick popularized by anglers fishing at night is applying a small amount of rigging thread or mono to a specific section of your fly line. By having a marked-off length of line and a small piece of material notched to it, you can have a tactile indicator when you've hit that length. The line will "tick" your guides, thus offering a physical signal.

Fish smaller rigs. Not necessarily smaller flies or tippets, but by fishing single-fly rigs, you can eliminate the potential for tangles or buggered-up flies. With limited visibility, it's not always easy to catch when that dropper fly tangles up or snags some part of the larger fly. Keep it simple to fish and simple to replace should you need a different fly.

Hoot Owl fishing restrictions prohibit fishing between 2:00 p.m. and midnight. While this issue has occurred mostly in Montana, other Rocky Mountain and West Coast states have enacted similar nonfishing times. Fishing closures prohibit any angling activities on the stretch of water specified. You will never study water harder than when you are on it and not allowed to pick up a rod. We once spent a few days floating on some of our favorite water with no rods in the boat. The water was open and almost nobody was on it. We studied drifts we had not previously fished and learned a lot about the water we have fished hundreds of times.

Take advantage of any time you are allotted between work, family, and whatever. On a larger scale, think about the fish. Even if the water is open for fishing, think seriously about the impact you may have on the fish. When the water creeps to 72 degrees, the mortality rate of catch-and-release trout skyrockets.

Fish and game departments are not always the quickest and most reliable response organizations. While not putting a massive "Please target me" sticker on our backs, it's important to remember that these are state employees. Like a lot of us with bosses, the wheels of progress move slowly. Many times, the red tape gets in the way of quick and appropriate action. Now is when we need to point the finger at ourselves. I know, I know, it's hard to take any responsibility for our actions when we can just blame others.

However, some self-responsibility here can do a lot to shift the tides of when it's cool to haul a fish out of the water for a sweet Insta pic and when it's best to let the river and fish rest. Unlike the proposed electricity reductions many of us get asked to abide by in the summertime, take this seriously. Don't take the selfish point of view, "Why should I do it, when I know those people aren't?" It's best to find a local lake or other warmwater fishery that will not be negatively impacted by the presence of angling.

Metered Fishing Time

Like deciding when to "cut and run" when changing tangled flies, anglers should consider putting a stopwatch to their fishing.

When fishing seasons and restrictions affect opportunities and water time, thinking smaller can better optimize the effort. A beneficial tactic

Two anglers working river right, on a set clock to maximize their efforts on a wide, sweeping western river. By breaking down large rivers into bite-size, attackable sections, better fly presentation and more efficient use of angling time will occur. The cleverly positioned turkey talon reminds anglers to think small and fish smaller. PHOTO BY NICHOLAS CONKLIN

on heavily fished water is to put a mental "timer" on a specific portion of water. Being the first on the water, at the launch, is not always the best use of time and energy.

There is a particularly interesting-looking bridge along the Upper Madison River that, while popular during daytime floats, often gets overlooked during twilight hours. While most boats will briskly slide underneath the bridge or use its corners as a lunch or bathroom stop, they seldom spend time fishing the deeper runs that have naturally formed behind the bridge pilings. Additionally, this manufactured structure has helped shape some deep cutouts along the bank.

While not unlike many other wooden bridges that crisscross the rivers of the West, this bridge allows for some varied bottom depths and fish-holding lanes behind pylons, in total a 40-yard stretch that can be fished within about a 30-minute window. It's high cut, and there is limited wading as you walk down toward where the bridge meets the land. High boulders on both sides add to the difficulty. There are only around 20 yards of land on each side that can be fished without hanging up on brush and getting into difficult casting angles.

This is a great situation in which a time-structured session can be done around sunset, with a single rod. While better suited for the wading angler due to time constraints, this could also be accomplished on a short or regular float. In this case, we will focus our example on the wading angler. Whether approached with a streamer, on the swing with a longer rod, or with a high-stick rig, this stretch is a fitting example of choosing some underfished water, approaching it with a single technique, and fishing it for several nights.

Easy to manage and distinguish tippet spools help keep the angler fishing within the stop clock. Color-coding of tippet spools, spare line spools, rod socks, and anything that can aid in efficiency should be utilized. These tricks will set you apart from other anglers fumbling for the right tippet or fly. PHOTO BY OLIVER SUTRO

Even some of the most popular and well-written-about stretches of trout water will have calm, unpressured mornings. This is when varying seasonal travels and consistent notes can help anglers approach less pressured fish.
PHOTO BY NICHOLAS CONKLIN

You'll become even more proficient if able to fish this same stretch throughout several seasons. You'll be surprised how holding patterns change based on water levels, available sunlight, and temperature shifts. If not, the focus should be short on time and with a technique you are comfortable with.

In these timed scenarios, avoid fly, sink tip, or tippet changes if possible. Keep your terminal tackle simple the first few trips out. As you fish this stretch more frequently, you can open the bag of tricks and begin to experiment. However, a single-rod, single-fly approach in a situation like this one can allow for a very pure, simplified fishing experience.

Often, dumbing down the approach will yield superior results. Choose a fly color or size based on your conditions and confidence, and fish it through the entire stretch.

If time allows, you can restart this short stretch with the same fly. Fish it through and then call it an evening. We like to call this a two-beer fish. Spend enough time to comfortably polish off two cans of your preferred beer, and then reel in and head to the vehicle.

It can be a straightforward way to break down a stretch of water, keep things simple, and be effective. This can be equally beneficial for those who have spent the day traveling and fishing different waters.

By adhering to self-imposed time constraints, anglers can better focus on goals when pursuing highly educated fish.

It's not always easy to be up and out the door before the sun peaks out from over a hill in the West. However, this early movement can help get anglers where they need to be, and on the water, before many others float or wade by. PHOTO BY CAPT. COLLINS D. ILLICH

CHAPTER 3

TIME DRAGS ON: FISHING WHEN OTHERS WON'T

Perhaps Roger Waters conveyed it more clearly when he posited about no longer preparing for life but being right in the middle of it.

As anglers, we often spend many nights dreaming of being on the water, whether it's a warm, saltwater flat serving as a distraction from wheezing winter winds that screech outside the window or the first warm breezes of spring that shift your mind to thoughts of the trout opener. One of the simple pleasures of fly fishing is just the amount of time we spend thinking about it. Now is the time to take those dreams and transform them into actual angling opportunities.

Much of this book discusses the aspects of preparation and how proper planning, evaluation, and adjustments can lead to better angling success. What we don't want to neglect is when you are right in the mix and need to go out and fish. There is likely a reason why traffic increases so much during certain times of the year on western rivers. Whether it is a hatch, improved weather, or anglers with time away from work and school, certain patternable days will have the most anglers hitting the water. What we'd like to encourage is the ability of the agile angler to choose the not-so-common hours, days, and weeks to find fish.

In forthcoming chapters, we will lay out thoughts and plans for becoming a more effective angler in highly pressured, heavily fished waters. A beginning stage to that goal is how these off or less obvious fishing times and seasons can build confidence. Without the prying eyes and heightened anxiety that busy water can generate, these slow times can provide a great warm-up, even for experienced anglers. By developing the techniques and tactics presented in this book, anglers can build their skill sets in these slower times, at a much more digestible pace. If working on some new casting ideas, trying out new flies, or casting a new line on a rod, these less pressured times are great for confidence-building while still affording anglers a real-life test. Don't underestimate the ability to try new things in relaxed environments.

In discussing the busy waterways and crowds that effectively make fly fishing the West more difficult, one of the most daunting but easily correctable issues is the physical approach. What is important here is that certain seasons and hours of the day will be more productive than others. Anglers who don't make the effort to fish, and fish in what would be deemed "off hours," aren't going to improve. While futzing around the fly-tying table or digging through gear is a great time killer, it's just bad foreplay. Intrepid, adventurous anglers will consider fishing at times that may seem outside of the norm. They will try different techniques, at times of the day that others may scoff at.

Either way, your time on the water and when you fish can have an important impact on your angling success. Depending on the season, water flow, and the previous winter's snowpack, you may be faced with some daily restrictions. In positive-water years, and with no mandatory restrictions, anglers still tend to gravitate toward the familiar.

Fishing Early

While there is always the temptation to spend another 20 minutes in bed (fending off the aftereffects from the night before), the courageous angler knows that the first chance at a run or a drift can be the best.

Early mornings are synonymous with fly fishing, but always being the first up isn't the answer to success. Even though there is a desire to be first to the ramp or parking lot, truly successful anglers prepare long before the alarm sounds and the coffee starts percolating. PHOTO BY NICHOLAS CONKLIN

Off hours and days, poor weather, and popular holidays can provide a variety of open water opportunities for the flexible angler. A snowy creek in a popular ski town provided the perfect morning for tossing a fly.
PHOTO BY CAPT. COLLINS D. ILLICH

Most western boat ramps in early spring are ghost towns until 10:00 a.m. The likely culprit is the stereotype of water warming. Perhaps the snuggly mountain towns of Colorado are still dealing with melting snowpack and the water is less than desirable, or the technique of the day calls for the sun to warm the water and for the fish to "wake up." Whatever the excuse, these ideas have taken an all-too-important hold on angler behavior.

Anglers seeking the advantage of lighter crowds and less pressured fish should always consider approaching early season fishing in the early morning hours. Tossing out modern convention can accomplish a few things. First is the mental aspect of getting to the ramp or wade access spot as the first vehicle. Whether fishing with a group or solo, there is almost always a sense of urgency as you hustle to the fishing spot. This issue is amplified when behind schedule. Like a desperado waiting for a train, we've likely all had that clock-watching, anticipatory feeling when waiting on a ride from a fellow angler. It seems the driver hits every red light, and every step in the adventure takes just a little bit longer. The sun begins creeping up, and so does the anxiety.

We recommend arriving extra early at least a few days a year to combat that feeling. Whether fully intending to step into the river before sunrise or simply as an exercise to reduce transit stress, this method can allow an angler to transition into fishing mode more peacefully and mentally. For those who still struggle with pre-trip preparation or simply haven't been able to commit the time, this approach can allow for a less rushed time to assemble rods, rig flies, and start the day. It can also allow your fly to be the first one through a given stretch of water.

While we attest that "physical" traffic (i.e., boots in the water and boats floating by) plays a role in the overall pressure a system faces over the day, the flies tighten the screws. Whether effectively placed or haphazardly slopped and ripped across the water, every time a fly hits the water, it heightens the awareness of trout. The ability to be the first angler through the water can give you the best chance at a hookup.

The early season confidence will be a key building block to success in future trips. Beyond fishing early in the day, there are obvious benefits to shifting the calendar timing when approaching a specific river.

Instead of waiting for the popular late spring and summer bite, anglers should consider the "off months"—early spring, late fall, and winter—when trying to separate from the pack. Pending local restrictions and water closures, willing fish can often be found during these less-than-pleasant weather times. While the blue-haired, slow drivers in the sunset of their lives may pack their Cadillacs and Buicks and head for warmer climates, western trout aren't so lucky. Though feeding and holding patterns may change, as will their location in a river, fish can't go far and still need to feed. One way to chart success and potential opportunities during these periods is to consider fly choice.

Like the undulating, snakelike curves of the western landscape, we feel that the approach should follow similar writhing movements. Fly selection should change and evolve due to conditions. The early morning bite can be further exaggerated by encouraging an anger-based strike from a fish that may be somewhat lethargic. A simple meal of a small nymph within proximity may be appealing, but it might take something substantial to get a fish to expend the energy—similar to the transition to fall, when fish begin to sense the oncoming changes in the barometer and are eager to hunt down larger, calorie-packed meals. As the day wears on and more bug life arrives, the need to chase down one big meal

Despite the hardpack snow and overcast day, Nicholas found plenty of open water and willing fish. This very late season fishing in the Northern Rockies can provide some serene, fun outings. PHOTO BY OLIVER SUTRO

High water and aggressive rapids aren't the end of fishing. Anglers should think about fishing tight to boulders and other obstructions. Look for deeper pools and places where fish can hide out when the water gets high and muddy. PHOTO BY CAPT. COLLINS D. ILLICH

A late season float on a sunless day has allowed these two anglers to work both sides of the river. The limited traffic was possible due to these anglers rehashing notes and reports from previous seasons. PHOTO BY OLIVER SUTRO

diminishes, and most trout will move to smaller food sources.

As the snow melts and inundates local waters, approach them aggressively with larger patterns and brighter colors. Bemoaning a "blown-out" river only adds to the negative and challenging stereotypes that engulf the sport. While fish will be pushed around the system and be in vastly different holding lines come July, it can be a time to experiment with more aggressive patterns. The overall pressure will likely be less, but the amount of water and pressure from nature will require an enticing fly. These high, off-colored flows aren't always the best for easy-to-spot bug activity, so we move away from small precise patterns here.

To encourage movement, we tend to lean toward larger, gaudier patterns. We don't, however, abandon the theory that these trout will still seek the path of least resistance. They may be tighter to the bank, and they may also seek out the smaller channels, back eddies, and other areas away from the most aggressive pushes of current. Fish are still in the system—they just likely have moved to a more challenging holding position.

Fishing Late

As spring transitions to summer and into early fall, the parking lots are packed, and you better have your act together early.

This time of year is typically more hectic, as summer vacations and domestic weather patterns have most anglers seeking the respite of the Rockies. With shorts and T-shirt weather comes the onslaught of angler traffic, as the weather patterns and typical hatch behavior bring out anglers en masse.

While you may have been able to ease into the day and not hit the launch until 10:00 a.m. back in April, these lots now resemble a Costco warehouse by 7:00 a.m. This period of increased activity can be ideal for those who are patient enough to wait for the drift boats to finish their floats and the wade anglers to reel it in and head to dinner.

Focusing on late afternoon floats and odd times and dates can help keep you away from other anglers. It can also allow you to have a more comprehensive understanding of fish behavior, as changes in their patterns can be observed. While it will require patience and a more fluid schedule, there is

With a powerful fall Trico hatch coming off, angler Jason Randall is still precise about his casts, unlike the other anglers around him. Their constant movement and casting doesn't result in more hookups. PHOTO BY NICHOLAS CONKLIN

value to this thoughtful approach. It will take work to entice fish that have likely seen hundreds of flies drift past them, but success can be had as the sun dips toward the horizon.

This may also be the only approach in some western states if Hoot Owl–type restrictions are in place. These condition-based restrictions are typically instituted during midsummer and can extend into September. During these periods, fishing is allowed only from midnight to 2:00 p.m. (or other predetermined periods). The goal of these limited angling hours is to help mitigate the stress forced onto fish as oxygen levels in the water decrease. As air temperatures rise, water temperatures tend to continue to rise during the season. Without significant rain or water released from a dam, the river will not have the ability to cool. With very warm nights, there is unlikely to be a significant drop in temperature to cool the river.

Anglers should be very cautious when water temperatures rise above 65 to 70 degrees for several days in a row. While most resources and studies suggest the optimum feeding temperatures exist closer to 50 to 60 degrees, even these temps can begin to exert pressure and slow feeding behaviors and lead to increased mortality. These restrictions vary depending on the state and even specific river systems, so the onus is on the angler. Local fish and wildlife departments or fly shops are great places to confirm.

The transition in fly size and style here is also important. In periods of higher angler activity, it is helpful to consider downsizing and being particularly conscientious. As fish see more and more patterns, and usually the same patterns throughout the day and season, your selection becomes crucial. During these times, we not only downsize flies but also seek out patterns that may have a slightly more neutral tint. One of the most widely used flies in the West is some form of Zebra Midge, usually used as a secondary or trailer fly. While the summer months may see more black Zebra Midges than stadium seats see butts, that doesn't mean subtle tweaks can't be made to these common patterns.

When fishing a multi-fly rig in an area that is heavily trafficked, something as simple as a midge pattern without a silver bead head or a body wrap can provide just enough variance. A dark-colored bead head and a fully black body can provide a sufficiently different look. If this variation isn't available, I've found carrying a Sharpie-style marker can allow for simple, inexpensive tweaks while on the water. The same approach can be used by the streamer angler, who may have fished the "hot" Sparkle Minnow color all day, with marginal success. Altering the shade or tint can increase hookup rates. Again, the marker or some type of color-altering device can add just that little extra here. Keeping black, brown, tan, pink, and yellow markers on hand can allow for some minor adjustments.

As the day wears on, this can be the time to slowly experiment with increasing fly size. With the onset of shadows, cooler water, and low sun, fish activity will begin to pick up. This daily fluctuation in activity is no different as we shift seasonally into fall and winter patterns. This is typically when the streamer junkies emerge and begin hitting the water with flashy, feathery concoctions. This later season puts fish into serious feeding mode to pack on enough mass for the upcoming winter. This period of frenzied aggression and feeding calls for meatier patterns and is an opportunity for anglers to take risks in their offerings.

As winter sets in and the first snows fall, it is wise for anglers to shift their approach to smaller, more present offerings. Smaller nymphs and even some dries are the game in the winter months. While there is an opportunity for an unpressured approach, anglers do need to consider that it will take more effort and more focus to hook a fish.

As discussed in the previous chapter, all of these "off periods," or periods of frantic weather, can be great times not only to avoid traffic but also to look at a river a different way. While this idea may produce a head shake or side-eye emoji from most anglers, these off periods, times of high, bright sun or low water, can be great for observation. This is when the truly effective anglers will always outfish others. Premier guides are very successful on their rivers and streams because of the amount of time they spend on them. They see how the river changes, not just throughout the season but also each day.

During a recent casting lesson, many of the students asked how to become proficient in their local waters. They seemed to fixate on fly choice, leaders, and rod style. While casting and knot-tying skills are

Wading close in and tight to brush and obstructions not only is safer for the angler but also allows the angler to target fish holding tight to the bank. Several things can push a fish tight to the shore, but one thing certain to drive them out is the boots of a too-eager angler. PHOTO BY NICHOLAS CONKLIN

critical, those are steps that should be perfected long before stepping into a pair of waders. Where we feel an angler can go from sporadic success to mastering a stretch of water is time. Time spent watching and fishing during various times of the day and different seasons, weather patterns. Not only can an angler learn much about what is going on below the surface of the water, but the time spent walking fishing paths, without a rod in hand, will also provide an opportunity to discover new paths. More efficient time to fish effectively will increase, and by walking and observing the surroundings, anglers will likely find new holding lies and areas to cast a fly.

Fishing smaller in low light applies to more than just physical distances. Not necessarily smaller flies or tippets, but fishing single-fly rigs. With fewer flies, there will be fewer opportunities to tangle and lose flies. With limited visibility, it's not always easy to catch when that dropper fly tangles up or snags some part of the larger fly. Keep it simple to fish and simple to replace should you need a different fly.

Easy-to-maneuver lighting systems are also advisable when nighttime fishing. A light and comfortable headlamp, ideally designed in this century, is important. Points to consider when researching a headlamp are the overall lumens, light, color options, physical weight, and battery strength.

The average lumen range of fishing- or hiking-style headlamps is around 300 to 600. The lumen rating is effectively the amount of light given off by a bulb. The typical, in-home 60W lightbulb gives off around 800 lumens. An effective headlamp should provide enough coverage when walking dark, bushy terrain but also not be so overbearing as to affect your ability to retie or handle knots up close.

Lighting options, or the ability to switch from a soft white light to a red light, are valuable for those fishing hatches at night. The ability to transition to a red or dimmer light still allows for visibility but will limit glare. This quality can help aid in wading visibility and won't cast a glare on the water when it is time to move—a welcome addition when fishing close to other anglers. Other safety features such as a strobe or pulsing light can be welcome in case of emergency or, at the very least, provide some interesting lighting effects for the post-fishing parking lot dance party.

Overall weight of the headlamp should be a consideration, just like it should be when loading a hip pack or other fishing pack. This item will need to be worn on your head for several hours. The benefit of having a lightweight, easily packable lamp should be apparent. With many headlamps being targeted at trail runners and long-distance hikers, the availability of ultra-lightweight, well-designed options has only increased. This is a good opportunity to spend some time in one of these specialty running or hiking stores, to try on a few styles to see which will be most comfortable.

The final component is the battery power and running time the headlamp allows. While we would never insinuate that the well-prepared readers of this book would ever forget to bring along spare batteries, a lamp with dead batteries is like a boat with a hole in it. Most will take a standard AAA battery—simple enough to have some replacement batteries in your vehicle, but also common enough to find at any gas station or local store. Many of the mid-level options feature rechargeable batteries that can allow for a simple charging boost on the way to the river.

Handheld flashlights and other gadgets are not advisable. Often clunky and difficult to manage, these will usually eliminate the use of one hand, making any type of rigging or retying of flies impossible. You're more likely to drop the flashlight along the path or into the water long before it can be an effective tool. Smaller portable, battery-powered lanterns can be great options for lighting a base camp. Whether along an open bank or on the tailgate of a truck, these are great options for a wider beam of light and keeping your headlamp fully charged until it is time to head out.

Nontypical Days and Holidays

While the temptation may be to sleep in on Monday mornings, the hungry angler will seek out these off days, as most anglers are forced to work. These early week outings also allow the river some downtime from the pressure faced on both Saturday and Sunday. Family-centric holidays are also times when we always look to hit the water.

Certain holidays and seasons can be a bright red, flashing sign drawing traffic to the water. Due to time off of work or a break from school, it is important to evaluate just how open the water may be. Vacation periods and holidays like spring break, Easter weekend, and the day or two before Thanksgiving and Christmas can be opportune times to find peaceful, unmolested water.

Spring break, regardless of location, will often find families focused elsewhere. This is typically when families and those in the angling scene are focused on ski trips or warmer climates. Depending on the snowpack and weather, the waters in the West often get ignored. The attention shifts to snowy peaks or flip-flop vacations in the South. While physical traffic in town may cause some concern, it is a safe bet that the rivers and streams will be open.

Easter weekend should be a viable target for open water. Whether religious or not, this is typically a gathering time for families focused on a nice Sunday dinner, not a fishing trip. This holiday of sorts is also the time when most homeowners will be drawn to outdoor gardening and home-type projects. While others are attacking the spring cleaning lists and overgrown landscaping, anglers should be on the water casting a fly.

For the more popular holidays, some discretion needs to be observed, as these are dates when it is important to be early and not late to the water. The days before both Thanksgiving and Christmas can provide some quality time on the water, hitting select runs before a rush of post-holiday activity. The day or two before Thanksgiving can provide for some excellent early fall fishing in the West. Techniques vary based on weather, but the likelihood of running into a crowd is minimal. Again, the advantageous angler can take advantage of others' last-minute runs to the grocery stores and planning for arriving family members.

The same can be said of the day or two before Christmas. As last-minute shopping, errands, or travel distract most of society, water is often left open. However, much past the initial hours of Christmas morning will find a flurry of activity. Whether it's the additional time off from work from now until the new year, a new tackle itching to be scratched, or just

Under a gondola full of skiers enjoying a spring break vacation, this angler takes advantage of a small creek in a Colorado mountain town. Although the middle of the day and during a busy tourist week, he was able to fish solo and cast to eager fish. PHOTO BY CAPT. COLLINS D. ILLICH

the need to escape overwhelming family members, Christmas Day will see the crowds return. Those with the time and patience to wait out the holiday rush can find open sections of water (if the ice stays away) in the days following the new year.

Being Prepared

A recurring theme throughout this book is the emphasis on preparation. Whether or not any of our other tips and tricks make it through the mental shield that inhibits the retention of information such as this, we hope that a deeper focus on preparation ekes through. Specific to this chapter, anglers wishing to approach the nonstandard hours, days, weeks, and seasons can always benefit from preparing a weather "go-kit."

Likely not much different from most gear, we suggest organizing some foul-weather gear, which can be easily grabbed and packed. The clever reader can insert any number of quips about how quickly the weather can change in the West. However, having an extra layer of clothing, a pair of gloves, or just a dry pair of socks on hand can extend an otherwise spent day.

Whether floating the Gunnison, wading the rivers east of Salt Lake, or waiting on a specific hatch on the Missouri, we have started many days cold and wet, only to have a bright burst of sun and warmth. Often, it has gone the other way. And the only thing that kept us on the water, not seeking a quick end to the float and a rush to the truck, was having extra clothing. No different from being prepared with a backup rod or a spare spool of tippet material, these "off" hours can be extended by having the personal gear to be comfortable. Comfort can lead to success. At the very least, it can help keep an angler warm, dry, and happy.

Your authors pictured were cold and wet, even with proper gear. This day started like many others, with a warm sun and the pretense of a comfy float. When the skies darkened and the rain came, we quickly shifted to the *American Gothic* pose. The backup jackets in the drift boat made for a safe, slightly less cold day. PHOTO BY OLIVER SUTRO

Various boxes for different times of the year can be helpful as opportunities arise with limited time to prep. This is a great time to downsize your most successful patterns and have them cataloged for various seasons and techniques.
PHOTO BY OLIVER SUTRO

Well-organized and packed tackle is another element of success. As conditions change, one of the most frustrating, time-consuming activities is the last-minute rush to grab gear. When all the chores are done and the checklist has been slashed, a spur-of-the-moment trip to the river is one of the many joys of being a fly angler. However, pulling and packing gear at the last minute is fraught with issues. Critical tackle items will be forgotten, and unnecessary gear is likely to be grabbed in haste. While not every angler will have the ability or storage space to prepare for every eventuality, there are some keys to be ready to roll in these nontypical fishing scenarios.

As seasons transition, we like to dive into the two or three rods and reels we most likely will need. By considering the two or three approaches you may need to cover the techniques in each month, you can assemble a go-kit. For example, November can be an incredibly fickle month. Many times, we have finished days around Halloween and into the first week of November fishing in a T-shirt. While most assume this is the ideal streamer season, the fish often don't agree. Apparently the dichotomy of throwing big flashy flies while enjoying the sun and a potential tan doesn't quite jibe with a big brown trout.

Instead of only gearing up with the heavier 6- and 7-weights, we have begun to pack a nymph-style rod as well. This way, not only can you cover a multi-fly, indicator approach, but, should a terrestrial bite occur, you can also smoothly transition to a dry-dropper setup: two rods, three techniques. A seemingly obvious float setup, anglers often develop a myopic approach, given their history of success and failures during specific months.

As a wading angler, it's hard not to leave the house without some kind of high-stick or Euro nymph-style rod. Jokes and aggressive social media posts aside, an angler who demeans certain techniques but laments a lack of success need only look inward. While this is a leisure-time activity, and no one should be bullied into or away from a specific style, these multiple styles flourished for one simple reason: success. Others have come and gone. The return rate is what cements a style or technique into the fly-fishing lexicon.

For anglers utilizing smaller packs or bags, this is a great exercise to continuously reexamine gear and maintain a simple pack with all necessary flies, tools, and leader material ready to go. We also find multi-rod-and-reel carriers can be a great way to keep these rods and reels within easy reach. This allows for a cleanly packed and safe dual rod setup. It can be easily stored and labeled for when specific tools are needed.

Reserve clothing such as a spare rain jacket, sweater, hat, and gloves are other simple items that can be packed separately and at the ready. Despite the thrift store poaching by hipster youth and themed partygoers, a quality wool sweater should be in every angler's bag. Providing true warmth and moisture wicking, this relatively inexpensive option is perfect to pack away and forget about until needed. While there is no fault in spending big bucks on a virgin Scottish lamb's wool sweater from a trendy European brand, this likely isn't one you'd want to stash in a bag all year. The temptation to remove and use it during normal life is too high, and, like most items, it seldom makes it back to the place it was intended for. Ever lose a screwdriver? Same theory. When items get moved and not put in the proper place immediately after use, they seldom make it back.

Socks are another item worth always having at the ready. Whether to battle the cold or just as a dry fresh pair after a day of wet wading, there is no reason not to have a pair put away in a waterproof ziplock bag.

The final, and often forgotten, item is a good hat—not the trucker-style caps with every fly shop from here to Kamchatka branded on the crown, but a legitimate beanie or other style of winter cap. In the late 1950s, US Army manuals stressed proper headwear, citing the amount of heat that is lost from the human head. Like most things in modern

A simple wool beanie, a backup jacket, and a second pair of sunglasses can keep an angler on the water and happy longer in foul conditions. Simple, yet often overlooked, items are what anglers should be spending money on.
PHOTO BY NICHOLAS CONKLIN

By designating set tackle and clothing items, anglers can more effectively approach less crowded and less pressured days with success. PHOTO BY NICHOLAS CONKLIN

society, that theory has been heavily contested, with some scientists and researchers suggesting that just as much (if not more) heat would be lost by someone not wearing pants. While we don't intend to sit around a lab discussing scientific matters such as this, we will contend that if you are lost and without pants, things have likely taken a disastrous turn. Or, at the very least, you've reverted to your college days. It's likely a story for another setting.

We feel that a proper hat can add hours to the angling day. In our case, it's about comfort and enjoyment, not pure survival. Like gloves, this simple item is one that most folks have several of, but they can find strange places around the home and vehicles to hide. So, designating a few hats, in easy-to-find locations (or, better, packed within your fly gear), will ensure you have one when that warm, 70-degree day turns cloudy and snow begins to fall.

By designating set tackle and clothing items, anglers can more effectively approach these typically less crowded and less pressured days with success. These steps can help eliminate the last-minute rush to get to the water, while ensuring the right gear to survive through a day of variable conditions.

One benefit to fishing in crowded water is the ability to watch and analyze what the successful anglers are doing. Not only does it allow a different angle to see what is (or is not) working relative to your position, but it can also dictate when it is time to call an audible in fly choice. PHOTO BY OLIVER SUTRO

CHAPTER 4

PAYING ATTENTION

In chapter 3, we discussed how skilled anglers can precisely dissect both heavily fished and often ignored secondary water. Anglers doing this enjoy much more success. The ability to understand fluid dynamics and fish behavior when observing water is a critical step to rising as an angler.

We intentionally posed the thesis from two angles to highlight how anglers can successfully fish the most popular water. It is critical to develop an approach when targeting ignored, overlooked secondary water. When faced with water that can be deemed to be "easy," or too "close" to the parking lot, or too high or difficult to maneuver, the entrepreneurial angler can pluck success from the jaws of defeat.

Now that target areas are acquired, the focus should shift inward, and the physical movements of the fly fisher can now be analyzed. How we as anglers physically move and the efficiency in our steps and casts is typically overlooked because we feel the more we cast, the faster we move, or the more water we cover, the better chance we have.

That's not always accurate. While the axiom that the fish are in the water so that's where your fly needs to be is true, we stress a focus on body movement and the expense of energy.

In this chapter, we discuss how personal movements offer a synopsis of the day's fishing, and how watching other anglers move while fishing the same water often delivers the full story. How successful anglers position themselves and how they move while fishing should be critiqued, and all their nuanced motions should be compared to your own. Much is learned by making mental notes of the physical movements, angles of approach, and how other anglers navigate while fishing. It is important to have some alternate idea or image you can overlay onto the water to help drive focus. Like NFL legend John Madden and his telestrator, develop some internal ways in which to catalog the movements of fellow anglers and boaters.

In fly fishing, we spend an enormous amount of time looking at each other. We study fellow anglers' rods, reels, clothes, and flies. We devote entire convention centers, websites, and pages of books to examine and slightly tweak what the person next to us is purchasing and wearing. We take casting lessons and deride others as they flail around a casting pond. But we seldom try to understand what they are doing, why they believe they need to move their hand that way, or why their rod tracks one way or another.

Humans are freakishly visual but lack retention skills. Blame it on the proliferation of the 24-hour news cycle. We demand constant visuals, but when the running ticker at the bottom changes, we've all but forgotten about the genocide in Rwanda in favor of the tropical storm engulfing the Lesser Antilles. We like to stare, but we don't really know why, and we seldom remember what we were staring at in the first place. Oh, wait, who is Jennifer Lopez dating now?

We look, but we do not recall.

At some point, we became more obsessed with the look and style and forgot to focus on what these things do, why an angler does what they do, and how it affects fishing success. While we discuss the more gear-associated success and failures of anglers in later chapters, our focus here is on the physical movements of anglers.

As colleagues in the fishing space, we've been lucky and able, for over thirty combined years, to spend many weeks traveling the world representing various factions of the fishing industry—working consumer and industry shows and dealer events, conducting casting programs, hosting fishing trips, and interacting with anglers from around the world. From the person who has never picked up a fly rod to the leather-skinned, raccoon-eyed angler who has been everywhere and thrown a fly at everything, one unifying factor among all skill levels, ages, and genders is casting.

We've been equally lucky to learn from and have spent personal time around some of the legends of the sport. Casters like Ed Jaworowski, Lefty Kreh, Bob Clouser, and Simon Gawesworth aren't mythical beings beyond the realm of the common angler. While immensely committed to the craft, they have

Success can be found on pressured waters, even within a foot of the bank. This willing rainbow was holding tight to the bank, awaiting some easy calories. While not far from the parking lot, many anglers who came before either fished too far out or flat out ignored this stretch of water. PHOTO BY OLIVER SUTRO

Anglers fishing together and watching others cast can be vital to skill building. Even if it is your home water, right off the steps of your front porch, watching someone new approach a familiar stretch will provide some fresh insight.
PHOTO BY NICHOLAS CONKLIN

developed because of a deep desire to better understand the end goal. They also understand that asking questions and learning from others is the best way to improve. If you take the time to seek them out at a trade show, they will share their knowledge with you.

But there are thousands of anonymous expert anglers out there. Each has modified their cast and skill set in one way or another. Most have shared all they could. For the most part, everyone in fly fishing is generous and genuinely interested in others' success. Most did not find the notoriety of casters like Ed, Lefty, Clouser, and Simon, but they all made contributions to people's better understanding of the art and skill of a cast.

Outside of the hours of stories, jokes, and cheesy one-liners, one of the most valuable things learned by watching them is that an angler's skill and willingness to learn will ultimately impact their effectiveness. Like not being able to buy the skills off of a rod rack, an angler must not mistake the age or time spent performing the task as competence. Like a demented game of Six Degrees of Kevin Bacon, watching a caster and understanding how they respond to the advice and pointers of a skilled teacher will aid in the angler's evolution. Claiming to have read and studied the greats, while overanalyzing and dissecting their approaches, typically won't bring a caster to the next level. Instead, you should take bits and pieces from each person and find where you can mold those principles into your own style.

This is where the fishing "buddy" becomes invaluable. While they may be great to have around to split a tank of gas or bring the beer, their value lies in more profound areas. By providing a second set of eyes, many casting improvements can be made. What the angler performing a roll cast or working a double haul may not catch amid the action, a partner a few feet away may notice. This is the time and place to take the pre-fishing practice tips covered in chapters 5 and 6 and apply them to what others on the water are doing. Or, more important, you will discover what is not being done and truly learn why other anglers are not as successful.

If you have read this far, you are extending some semblance of credibility to your authors. Before we burn any lingering credibility, let's discuss some of our casting theories.

Casting Proficiency

Casting proficiency is the efficiency of the movements an angler makes during the cast and the energy expended while making those movements. Without the proper foundation of a cast and the skills to adjust the style of the cast based on the technique and surroundings, success becomes unattainable and frustration creeps into the angler's thoughts.

In this chapter, we discuss the value of practice and developing strong foundational casting approaches before a trip or a season. This will often depend on the anticipated techniques needed for that style of fishing. Not all casts are created equal, and no one style will work in all conditions. Repetition here is key, as is the ability to rebuild or restart the cast when issues flare up and things go awry.

An angler headed for their first saltwater trip needs to master a separate set of skills from those of an angler approaching a technical spring creek. Thoughtfully analyze the anticipated skill set required for success for your upcoming trip and/or season and put in the time to be able to make those casts until you cannot perform them wrong.

Regardless of technique, time of year, style of fly, or type of rod, experienced anglers still need to master some basic steps to build an effective cast. While casting solo is effective at knocking off the rust and getting timing back, group settings and lessons with qualified instructors are a valuable part of the toolkit.

By working with others, watching others cast, and soliciting the opinions and ideas of those in the teaching realm, small tweaks can be adopted to further efficiency. Many talented groups exist, such as Fly Fishers

Casting proficiency does not come overnight. Working with a proper instructor and understanding your physical skills and limits will make the entire fishing experience more fruitful. PHOTO BY OLIVER SUTRO

Without the proper amount of line and mass moving in the backcast, the forward cast will struggle. The caster here has the right amount of mass moving behind her, putting a nice bend in the rod. PHOTO BY NICHOLAS CONKLIN

International (FFI), along with local fly shop resources that can help refresh and fine-tune casting strokes. Many such guides, schools, and skill improvement programs exist. Seek them out and learn.

From thousands of hours of angler observations, we have concluded that many of these casting hiccups can be broken down into four phases, regardless of skill, experience level, or type of tackle. To keep it simple, we have provided these four ideas with a helpful mnemonic device: L.A.S.T.

The L.A.S.T. Method

The L.A.S.T. method focuses on the four base principles of an effective single- or two-handed cast: **l**oad, **a**ngle, **s**troke length, and **t**empo. Simple enough to remember at the end of a long, fruitless day of fishing when casting faults creep in and issues often compound on themselves.

Each of these steps is vital to forming a complete cast. These rules are as important for a new angler looking to build a foundation as they are for the more seasoned angler looking to correct. Both can utilize these ideas when form falters and casting faults emerge. This four-letter device will allow anglers to analyze, reset, and rebuild their stroke to return to effective fishing.

The first phase is the **load**. Loading a fly rod is an overly complex description. It means to "bend" the rod. The bend is created by resistance. Most often resistance is created by the mass of the fly line accelerating away from the rod. This critical step must happen before progressing into the upcoming cast's next steps. The rod must be loaded before moving into the forward cast; then the rod will transfer energy from the loaded rod to the fly line, leader, tippet, and finally the fly.

The **angle** or path in which your rod tip travels during the cast is your next consideration. Your fly line is going to follow a loaded rod. Specifically, it is going to follow the tip of the loaded rod. If you are frustrated with your cast and you are indeed loading the rod, consider the path of your tip. Remember that what the rod tip does, the fly line will do. The fly line can function as a great flashing highway sign telling you what type of path and movement your rod tip makes.

The angle or path in which your rod tip travels during the cast will determine how the fly line unrolls. The word *cast* is almost a misnomer here, and considering *unrolling* a fly line can lead to a gentler approach. PHOTO BY OLIVER SUTRO

The **stroke length**, or the path along which your body and rod must travel, is dependent on the conditions (length of line and leaders, water size, or targeted distance). This is a dynamic component of the cast. While where the stroke length starts and where it ends may change based on the above considerations, anglers must remember that the distance targeted and length of line needed will change the stroke length.

One of our favorite analogies from author, teacher, and casting authority Ed Jaworowski is to imagine yourself throwing a baseball. First, imagine being 10 feet away, and then 30 feet away, and lastly 60 feet away. Think about how your body would adjust to the distance and how your throwing arm's motion would change. The same applies to casting distance and control.

The final phase is the **tempo**, speed, and acceleration of the cast. *Acceleration* is derived from a Latin word meaning "to hasten" or "to speed up." This final step, while the last in the chain, is as important as the previous three due to one of the more common results. The tailing loop (unfortunately misidentified as a "wind knot") has a high likelihood of resulting from the erratic or inconsistent implementation of speed. For more experienced anglers, this idea is absolutely one you should keep in mind as the day wears on and your effectiveness suffers. It happens to everyone. While you may start the morning off with a nice casual cast and swing, the day drags on, oftentimes exacerbated by limited strikes and fewer landed fish. For new or intermediate anglers, this idea can be applied to form the foundations of the cast.

One of the most common things anglers do when a cast starts to fail is to try to power through it. Let's add a bunch of power and muscle to it—that'll work! Power isn't necessarily your friend with a fly rod, or, better said, haphazard and inconsistent application of power is not your friend. All the power in the world is not going to load a fly rod if there is no mass being put into it.

THE LOAD

The load stage with a single-handed rod can come in many forms, but it needs to be executed correctly, or else the rest of the cast will suffer. In that case, you will be sad and want to throw the rod in the truck and cry for a few minutes. . . . Despite what Robert Smith of the Cure says, boys do cry.

The most effective way to generate beautiful "candy cane" loops is to develop the proper bend in your fly rod. This is when the correct amount of line is aerialized and the fly line is pulled behind

The waterborne anchor will need to be pulled from the water after sufficient time to stick to the water. That stick, or line lying on the water, acts as the loading device, like how an aerialized cast would develop. PHOTO BY OLIVER SUTRO

the caster. Without the proper portion of the fly line heading out of the tip of the rod, it is more difficult to properly bend the rod, and thus the unrolling of the fly line and leader will not develop.

The same applies to any waterborne or sustained anchor type of cast, whether a simple single-handed roll cast, single-handed Spey, or two-handed cast. Without the bend, proper progression of the cast cannot occur.

For a single-handed aerialized cast, the load comes when the appropriate amount of fly line is outside of the tip of the rod, and its mass is pulling on the rod, or causing it to bend. For a waterborne cast, the anchor or loading phase is most effective when the head of the fly line grabs or sticks to the water. The anchor, or line sticking on the water (surface tension), is what generates the load, causing the rod to bend, which will allow the caster to unroll the line after the stop following the forward stroke.

Load can be inconsistent when there is not enough head or line on the water or only a portion of the head or line lands on the water, thus not allowing for an appropriate amount of mass to build (waterborne anchor cast). The same issue arises when an ineffective portion of the head is outside of the rod tip during an aerialized cast.

For most anglers, this problem will occur with a single-hand cast if there is not enough fly line out of the tip, which typically results in an awkward or forced cast. Too much arm is imparted into the forward cast, often pushing the small amount of weight (fly line) forward. Or there is not enough time for the head to "stick," or grab the water, and the angler moves too quickly into the lift and sweep.

The development of short-headed fly lines has both aided and hindered casters as they develop components of the effective cast. While they have significantly opened new water and afforded anglers a more efficient system to pick up, deliver, and fish larger flies and lines, in more compact spaces they have equally aided in casting faults. Because these lines tend to have heads ranging from 11 to 16 feet, anglers often do not take the time for the line to sit on the water or move through the backcast. This again creates a lack of load.

Bear in mind that despite the thousands of dollars anglers spend on rods and lines, and the countless hours that go into the development of these items, the object is to simply bend a lever and unroll a line. Making a fly rod work correctly is remarkably simple physics. Be damned composite materials and nanofibers, with your high-tech carbon layup and space-age guides!

THE ANGLE

Once you have gone back through the load phase and examined it given the target and objective, you have the right amount of mass to bend the rod. The next step in the process to consider is the angle of the rod tip and how it is moving during the entirety of the cast.

If you take anything away from this chapter, or any of the writings of your prolific casting instructors, it should be to always be conscious of what the rod tip does throughout the cast.

During an aerialized cast, anglers should adjust the path the rod tip travels relative to the horizon. Whether the object is to keep a low angle to cut through the wind or to keep a low profile and shoot a cast into tight cover, keeping the rod tip on a consistent path is vital. With a longer rod and line, however, this task isn't so easy. It is something that happens during a fishing day.

Never forget that what the rod tip does, your fly line will do.

Your line can be a great detector or indicator of what type of movement your rod tip makes during the backward and forward cast. As the rod tracks during the sweep or the forward cast, if it does not maintain a consistent path or it jumps, leaps, or wobbles, what lands at the end of the cast will be the result. The line will pile in a heap or land with a curve or snakelike undulation.

This is not to suggest that a rod tip cannot start at one angle or position and end in another. That is an important distinction. What is critical is the awareness of the movements the rod tip makes during that path. As a general physical theory, and with some exceptions, the straighter and truer your rod tip path, the more efficient and accurate your cast. Anytime the rod tip moves or deviates from its path, it can cause shock waves. These shock waves may be visible when the line lands on the water, but what it really does is create slack. In most scenarios, that slack will cause a loss of energy, which is not beneficial. Again, there are no set rules here, but whatever arc the cast makes, the tip should continue on a consistent path.

Whether shooting line under a strong breeze or keeping away from obstructions from behind, anglers need to focus on a consistent path. Using static objects like where the bank meets the tree line or the horizon can help guide the rod tip along a straight path. PHOTO BY OLIVER SUTRO

Often the tip of the rod will dip or move outside of the path or plane, which will cause a fault. These subtle, often unnoticeable deviations in movement can impart slack. When slack enters a cast, it lessens or weakens the amount of mass being generated to load the rod. Without mass, the rod will not load, the fly line won't have the energy to unroll, and the fly will not end up at the target. This problem is often common with new and intermediate anglers, not only during the forward cast but also during the initial lift and move to the backcast.

Ed Jaworowski frequently recounts the struggles he endured when he first began casting a fly rod. His instructor at the time did not allow him to make a forward cast until he perfected the backcast. He made ten thousand backcasts in his street before considering his forward stroke. While most anglers focus solely on the forward stroke, Ed recalls how his casting instructor stressed the importance of the backcast and being able to critically analyze the whys and hows of the rod tip and fly line movement.

After thousands of backcasts in the dim streetlights of his neighborhood, Ed more than mastered the backcast. He shares that wisdom with his new and even experienced students when they begin to have tunnel vision relative to their forward cast. What most anglers ignore and fail to acknowledge behind them is critical to accomplishing a positive forward goal.

A common casting fault is when the rod tip dips or is even aggressively thrown downward during the backward movement. This issue stems from and can be corrected by monitoring the starting and stopping of the rod tip. Casters often develop tunnel vision and spend almost no time watching their backcast. This is something that should often be done in stages. While developing casting skills, anglers should focus on watching their backcast, with their eyes looking over their shoulder at the backcast. Once they can make that backcast every time without getting it wrong, they should focus on making that same cast while looking forward. By piecing these phases together, the entire cast can be performed. Waiting to get to the water, to spend the morning staring behind you, will not get you into an active, forward-facing fishing position.

All of these faults typically are a result of a tired or lazy angler. The angle stays too low or flat and creates more of a dragging V-shape than a D-loop. While this can be better performed with a longer lighter line, heavier shorter heads with sink tips won't allow for the proper backcast if anglers cannot keep steady rod tip elevation, thus affecting the forward cast. If the rod tip does not track forward in a steady line, slack is introduced into the cast.

To keep a steady forward stroke and smooth rod tip movement, consider utilizing your surroundings. One way in which the thoughtful angler can refine rod tip movements to rise above a nearby caster fishing in front of or behind them is to utilize fixed points on the horizon as an aide. Isolating a tree line, bridge, roof of a barn, or other straight object is a wonderful way to trace or have your rod tip follow along. By a turn of the head and shoulders, you can begin to redevelop a proper path for the tip.

Anglers need to add this point to their pre-trip practice routine. By performing some "maintenance" work on the cast before stepping into the river, anglers can move quickly through this stage and on to active fishing.

One trick to refining rod tip movement is utilizing the double haul. While this has been presented as an exercise to unite the movements of bending the rod and hauling line, it is ideal for a refresher in rod tip movement. While the focus is typically on the counterhand motions needed to carry more line, instead focus on the path of the rod tip. The utilization of a long length of colored rope set perpendicular to the caster presents a straight, fixed point that the rod tip can trace

If the rod tip does not track forward in a steady line, slack is introduced into the cast. Fear the slack! PHOTO BY OLIVER SUTRO

along. When warmed up, the angler should begin to lift the rod and experiment at different heights from the ground. As the rod elevates, the angler should also consider the V-shape motion most common in the casting stroke. While often used for new anglers, this image can offer a more effective tool to guide the starting and stopping position of the rod.

The addition of targets can then allow the angler to progress to a focus on stopping points. While tracing the rod tip along a straight path, casters should then move attention to the stop point. A simple hack to refine rod load and unload is the emphasis on the stopping point of the tip. This is often referred to by instructors as the "cone drill."

Drifting or not stopping the rod tip soon enough to form the D-loop, or at the stop at the end of the cast, will cause line issues. These line issues are great indicators of what your rod tip is doing. Too abrupt of a stop, and the result will terminate with a bungled pile of line at your feet. Too late of a stop will result in an aggressive dump of line on the water, with a loss of distance.

Again, what your rod tip does, your line will do.

These issues are common regardless of the style of rod or line system used. While rod length will play a role and may require some adjustments, a longer lever can lead to the same problems. Pileups, short casts, or lines that do not lay out evenly/straightly are great signs that the rod tip did not follow a steady path.

THE STROKE LENGTH

The stroke length is the third area of focus for anglers struggling to develop an effective cast. Stroke length is the motion or movements your hands and rod make during the cast.

This will depend on the type of line system or length of head you will be casting. Longer, more tapered lines typically demand a more elongated stroke. If your objective is stealth, attempting to cover distance and deliver a fly to a wary fish, you may need to exaggerate your stroke to properly load the rod with the longer-headed fly line.

A short, weighted streamer setup may not require a long stroke and path of the rod. Compact strokes

Short casts require short strokes. Longer casts do not efficiently originate from the rod hand; the energy comes from the hauling or line hand. A focus on pulling and following with the line hand for longer casts is important in the stroke length phase. PHOTO BY OLIVER SUTRO

It can help to photograph or video your casting stroke to analyze and dissect how much movement the rod tip makes relative to the path needed for an effective delivery. Another set of eyes is always helpful, as is a recording you can watch later. PHOTO BY OLIVER SUTRO

for line and fly scenarios such as this will help preserve energy and deliver just as effective outcomes. This type of tackle does not necessarily require a dramatic *Saturday Night Fever* dance move to reach targets. For those floating and hitting the banks, the shorter stroke can be better suited to the repetitive nature required by this approach.

The farther away from your body the handle of the rod is, the more likely issues will arise. Not only does the added weight stress the body, but it also leaves an open door for the first phase of this correction to occur. The farther away the rod is, the more difficult it is to manage, which can cause the rod tip to move unevenly. This is where the length of the lever is beneficial; one reason the 9-foot fly rod has become standard is due to its ability to move a line more efficiently.

While casting purists may bemoan the wide, sweeping, windshield-wiper casts often seen from western drift boats, they serve a vital function. Guides have learned and developed the technique allowing anglers to deliver a multi-fly indicator system. It also helps mitigate tangles and other time-consuming issues that keep flies out of the water. The more courageous rowers have taken to arming customers with 9½-foot and 10-foot fly rods. These longer levers allow for better repositioning of these longer-line, oddly weighted approaches.

The natural tendency of less-than-seasoned anglers, when confronted with this technique to deliver all this weight spread across a long leader, is to exaggerate the arm movement. Instead of keeping arms and hands close to the body and utilizing the length of the rod, they unconsciously reach out, feeling that the farther away the rod handle is, the more leverage and movement they will achieve. If anything, however, it creates more issues and inhibits efficiency. Thoughtful anglers should attempt to expend the least amount of energy while finding the appropriate stroke length to deliver the line and flies.

The same can be learned from anglers who don't attempt to adjust their stroke length as techniques and rigging change. The movement and path of the stroke must change between techniques. This is not a one-size-fits-all scenario.

Again, this step goes back to the path of the rod tip.

A precise motion with a tight loop to land a size 18 dry fly in a riffle is not going to be the best rod tip movement and loop size when casting a lead-eye streamer or multi-fly rig. The same can be said for anglers fishing larger flies and when sink tip or line density changes throughout a fishing session. The stroke length can go wonky when tips, mostly sinking tips, get switched throughout the day. You may become comfortable casting a short sink tip, but if you need to switch to something longer during the day, you need to compensate for that change. By adjusting the amount of lift when setting an anchor, and similarly when transitioning to the sweeping loop formation, the change in length and/or density will not have a negative impact on the cast.

Anglers should consider using natural and human-made objects to help track and correct the rod tip path during the cast. While they may be unsightly power lines or buildings, or natural tree lines and mountains, casters should use their surroundings to help rebuild the proper cast. PHOTO BY NICHOLAS CONKLIN

Wading depth will also change the necessary stroke length of a cast. While we'd all prefer to fish from consistent depths, much of what will be encountered during the day will change. It may be only a few inches, but it can lead to some stroke length challenges to correct. While in a boat, or off some sort of elevated platform, the floating angler will not have to make as many alterations as the wading angler. The likely changes to stroke will occur for the boat-bound angler as the width of the river changes or riverside obstructions arise. The focus then will shift to casting around, under, and in between these obstructions. These anticipated obstacles can equally be practiced during off-the-water time.

Anglers in both scenarios can achieve this result while keeping hands tight against the body and simply moving them closer to the head or, conversely, closer to the water. Move the rod butt and hands parallel, not perpendicular. This simple step can keep the hands within what we like to refer to as the casting box.

As hands and rod butts drift away from the body due to inconsistent or changing stroke length, anglers should imagine an invisible box sitting just off the front of their body. For the single-handed caster, the box should be more located to the side of the body, dependent on the dominant casting hand. If the box is open, and the caster is forced to move their hand into this confined zone, the length of the stroke, as well as the stopping point of the rod, will be further emphasized. This box size will change depending on the technique, but not during fishing. The point is to draw a line of demarcation, which hands and butts of fly rods aren't allowed to cross. This visual will help provide a mental image of where the most efficient motions should reside and what happens when that box is forced open.

You can also utilize this stage to keep an eye on hand movements and the stop. If the forward stroke drift occurs, here is where the angler can adjust the width of the box. Lessening the amount of internal space the rod butt can move will restrict the length of stroke.

The two-handed caster can also benefit from the casting box to fight rod drift. The most common drift that occurs with this type of tool comes from the rod handle extending too far from the body. Again, the amount of energy exerted here will build, and anglers will more quickly become tired and frustrated. Hands and arms do not need to extend in this system, despite the often longer rod levers upward of 11 or more feet. The functional positioning and repositioning of a line with a longer rod can be achieved within the casting box.

TEMPO

Tempo is the fourth component of the proper development of a cast. While it's the last letter in the mnemonic device, the astute reader will surmise the importance of all four phases to the foundation of a cast and the steps to correct it. When an angler struggles with the execution of one portion of the cast, the others will likely falter as well. Speed and pace are key in this stage.

The most obvious connection is with the load stage of the cast. When there is a failure to generate sufficient mass in the rod, the casting stroke is too short, or the angle is inconsistent, starting the cast at a higher rate of speed is not going to result in success. Applying power or pushing hard at the end of the cast will not compensate for the other incorrect steps. Anglers are far better off starting their cast slow and building up speed to a hard stop of the rod tip.

Author, presenter, and fly rod and line designer Simon Gawesworth has spoken often about the inconsistent speed of most anglers' rod movements. He references a car at a stop sign and asks the caster whether they want to be the asshole who slams on the gas, fishtailing, before tearing down the road, or, to get down the road as effectively as possible, do they want to smoothly accelerate while slowly building up speed? Of course, the dirty looks directed at the NASCAR wannabe are often the same garnered from fellow anglers while watching the high-speed attempts at a power cast. While you may not look as cool as the misguided caster who feels the need to add that extra push or back leg English, the result will be in the favor of the consistent energy-builder.

Inconsistent speed can occur for several reasons. One reason an angler rushes or alters their speed moving into the forward cast is due to an ineffective load. Additionally, an inconsistent rod tip angle may force the angler to rush forward, generating a false load.

As Simon instructs, one of the best ways to combat this issue is to start slow and finish fast. From the initial lift of the line off of the water to the top of the rod tip and unrolling line, speed should slowly and evenly build. Focusing on a steady, increasing tempo during the lift, to the backcast, and then on to the forward stroke will build energy and prevent any slack from entering the system.

If an angler starts the cast fast and powerful, slows down, and then tries to reinsert power or finish the cast strong, slack will be added. This inconsistent application of power not only creates slack but also limits the ability of the angler to generate rod load and line speed. This change in tempo also leads to the dreaded tailing loop. While some believe it's the fault of the wind, it simply boils down to the application of power and the path that the rod tip makes.

Muscles during this stage will tend to overpower the rod while ignoring the line hand. When performing a double haul or any type of cast at a distance, the line hand is what generates the power. What commonly happens when the angler does not have a consistent tempo with the rod hand is the line or hauling hand will also act inconsistently.

The L.A.S.T. method was developed not only to help anglers think more simply about the foundation of an effective cast but also to provide basic steps to rebuild a cast after a long, frustrating day. When considered and understood, the L.A.S.T. method will allow a reset and extend the day's enjoyment.

Lastly, practice all of this from as many positions as possible. The more varied your skills, the more likely your success. While standing in a field on a calm day is a great starting point, those looking to truly refine and develop real-world skills will practice in a variety of environments and conditions.

Now that the theory portion has been squared away, but before this book is slammed shut and eagerness causes a zombie-like barrel into the river like some crazed character in a Quentin Tarantino film, let's dive into ideas for observing fellow wading anglers. Evaluating movements, steps, and approaches to a piece of water should craft how you as an angler adjust to find more success.

After observing the potential holding lies from high above the bank, and having used the brush as cover, the angler stepped down the embankment and slowly worked her way upriver. By standing back and away from the water, anglers are less likely to either spook fish or miss possible targets. PHOTO BY NICHOLAS CONKLIN

The Approach of the Wading Angler

Every angler has been there. Our special piece of water, we know where we've had success and hooked fish. We may even have a strong inkling as to why we were successful.

But take a step back and examine the whys and hows of the way most anglers approach this section of river. How do they approach from their vehicle or the walking path? Do they sit and observe? Or do they make a beeline to a certain part of the pool or bubble line?

All of the high-pixel, fancy images in this book and other fishing publications cannot replace the fact that angling success and enjoyment can be determined before walking down to the water or slipping the boat off the trailer. If anglers continue to view and approach familiar water the same way, they are destined to have the same results. Anglers approaching unfamiliar water who rely too heavily on advice and visuals from others are also destined to reside in that purgatory of occasional success.

Awareness of surroundings often gets lost in the simple act of stringing up a fly rod and enjoying nature.

Nowhere was this the case more than on recent trips to a small feeder creek that leads into a southwest Montana lake. I (Nicholas) had scouted and fished this area during multiple seasons, at various water and light levels. While on the surface it presented as a very bear-laden, unfriendly environment, there had not been any sightings. Instead of keeping my eyes open, I spent many trips looking straight from the truck at the river some 30 yards below. With idyllic clear, small runs and gentle pools, I seldom saw cars parked alongside it, despite its location on a major roadway. Even with its brushy look, it was not difficult to find castable, reasonable water within a five-minute walk of the vehicle.

After several trips, my frustration grew faster than my limited success. Following countless fly changes, line tweaking, and all the normal countenance an angler should endure, I had resigned myself to abandoning the water. On my last hike back up to the truck, something caught my eye. Not 10 yards from the small overpass above the stream sat an eagle's nest. Perched high above the road, and with an equally attractive view of the water, it finally hit me. The eagle had a better view than I did, and the fish down below knew it.

What I had failed to notice in all those trips was that other anglers had noted and understood that regardless of the perfect fly, lifelike drift, and tight-looped cast, this was not safe water for trout.

POOLS

One of the more obvious and fishy-looking but frequently pressured water regimes is the pool. Typically deeper, slower water comes off an obstruction. Be it a grouping of boulders, large logjams, or other natural obstructions, pools can be ideal holding areas for fish.

The fish can sit deep or to the side of the pool, conserve energy, and easily scoop up nutrients tumbling down the river. Pools are the most obvious areas of least resistance, and anglers should be willing to work these deeper, slower waters with more than one technique. They also are an area where the patient angler can have as much or more success when following behind someone who has just fished it through.

But how?

Like the empty drive-through lane at a fast-food joint, the pool elicits the watery mouth and wide-eye gaze that going into a grocery store and cooking a meal doesn't garner.

While depth, speed, and overall river health are factors, the effective angler should approach this with more of an open mind than, say, a riffle or shallower boulder field. The deeper, darker, and slower water can afford fish a more substantial level of comfort. Although often targeted, pools usually aren't fished with as many techniques and for as long as they can maintain a bite. Because of this, anglers should focus on various body and fly placements, water depth, and fly depth.

The first focus should always be placement. This is a key theory throughout this book, and most other angling education, but is often forgotten in the heat of the moment. Often anglers forget the inside-out approach. They instantly target the bubbling, deep, plunging section of the pool and dump flies around the most turbulence. After a few drifts, they reconsider and start to toss a few casts higher up into the pool, when they are casting their flies up into the

A tight-line approach in the pool kept drag from entering the system and led to a hooked fish. The heavier, larger beaded flies and their ability to get down into the pool was key to a successful drift. PHOTO BY OLIVER SUTRO

obstruction, allowing for a higher percentage to snag boulders or other river debris. After a few casts and a few snags or general lack of fish feedback, they move on to another section, not understanding why they had zero success.

With a deep pool or plunging depth, anglers need to take the time to target the close-in water. Regardless of the size of the pool, fish tend to congregate around the edges. While the deep water may be alluring, anglers should consider working the slack water as attentively as the initial fast water at the head.

The initial approach should also consider the angler's physical location from the water. A slow and thoughtful approach at a pool should not differ too much here from a long open stretch. A metered approach also serves to enforce the ideal of casting short, working nearby water before extending fly placement.

As anglers move out and into the next portion of pool fishing, the overall weight of the rig becomes important as well. Because more water and a more aggressive change in currents can be found in this type of fishing, anglers must consider the type of water and flies they are fishing.

In this scenario, the current on the top of the water column is often moving much faster than the water at the bottom of the pool. What does this do to your flies?

If fishing an indicator-type rig, with a standard weight-forward fly line, the speed at which your fly line and then the indicator move will have a significant impact on how your flies move. As discussed earlier, what that bright indicator

Deeper, slower water and diverse pools should be picked apart, with anglers hitting the close water before working out line toward the opposite bank. PHOTO BY OLIVER SUTRO

is doing is not always what the flies are doing. This issue is further exaggerated in this type of water, as there is a greater variance in speed.

Author, presenter, and angler Jason Randall often discusses how most anglers assume their flies are moving at the same speed, but current speeds vary far more throughout the water column. If anything, multi-fly rigs or large indicators act more as drag indicators than strike detectors. So, the considerate angler should take the time to better understand current speeds and how fishing two flies isn't always the most appropriate course of action.

This is a water section where anglers could eliminate the drag-causing indicator and fly line and adopt a more tight-line or high-stick approach. Along with heavier flies, the thin diameter of the tippet and sighter material not only will allow the flies to cut into the water column, reaching a more appropriate depth, but also will afford the angler much more positive contact.

The same approach can be taken with a streamer or wet fly. Those wishing to swing or actively retrieve a fly through this type of water should consider a longer, thinner leader and a fly with more weight.

When we say more weight, we want to be clear that more weight doesn't mean more dressing or materials. While more heavily dressed flies can create more lifelike movement and cast a large silhouette, they also take more time to sink into the zone. Instead, focus on more sparsely tied streamers and those with large weights, such as bead heads and cone heads. These condensed, sparsely tied flies have a higher percentage of getting down into the strike zone and remaining there as the swing or retrieve is enacted.

The final phase in advancing when dissecting a deep pool is the ability to present flies from multiple angles.

Mobile anglers who take the time to work on both sides of the pool, present a fly at more than one angle, and more thoroughly cover the water column will outfish others. While it will take more time and energy to make the walk around to the other side, this extra effort can lead to increased success.

Anglers should consider working both the upstream and the downstream approaches to these waters.

Again, because of the depth and zone of safety, these pools can hold larger schools of fish, and they can often withstand more fish being hooked and pulled out of the pool as opposed to shallower-water holding areas. Nymphing anglers should approach both sides of the pool, either ahead or below, before switching sides and repeating the process. Before moving on to a different section of water, nymph anglers should consider retying with a jig-style nymph or streamer. Due to the location of the eye of the hook and the ability to animate the fly, anglers may be able to trigger a strike that the standard, drift-free drag did not accomplish.

Hookup rates and angling success are built on the foundation of thoughtful preparation and skill-building, which requires time and commitment to a body of water. This colored-up rainbow was landed after presenting multiple flies. It is a great case study of how most anglers would float this section, presenting a fly from only one angle. Here the angler was wading and swinging flies, providing a slightly different profile and angle. PHOTO BY NICHOLAS CONKLIN

Most nymphing anglers have had the concept of not moving the rod tip or adding any movement to their flies drilled into their heads. A change to a more active drift can increase the chances of success.

Flies like a Mayer's Mini Leech, Tom Starmack's (Zoetic Flies) Semi Seal Leech, or Sparky with a longer fibrous tail offer added movement and allow the angler to impart some jerks and subtle rod tip movements to bring the fly to life as it bounces off the bottom.

This is also an area where the swung or stripped fly angler can utilize the water for more unconventional approaches.

While the across-and-down approach is the most common, anglers should also consider a few aggressive casts and retrieves toward the head of the pool, with quick strips downriver. This can best be performed at the tail end of the pool, with the angler standing and casting at close to a 90-degree angle. After a few swings of the fly and the angler has worked through the pool, a different approach can be to cast up toward the drop-off, letting the fly sink at various intervals, before quickly stripping back down. This fleeing baitfish approach can replicate disturbed prey quickly moving out of the run and pushing back toward the end of the pool.

This technique could be especially useful if earlier fly retrieves did not elicit any strikes or a fish quickly flashed at the offering. This can push fish back, dropping deeper into the run. This method not only acts to congregate fish in a tighter zone but ideally angers them enough to strike at the fly when it is presented head-on.

POCKETWATER

Another water system that often gets fished sporadically and in peculiar ways is small to medium-size boulder fields, with small pockets of slower water. Often referred to as pocketwater, these clumped-together pockets of water can provide great holding locations for trout. Due to the energy expense required to move from pool to pool, these small pockets often act as stoplights.

Pocketwaters often contain a slew of small to medium-size boulders spread across a stretch of river. Oftentimes they are presented as loaded with obstacles and stretches where a fish needs to make serious movements to swim upriver.

Tiny pockets in this brushy New Mexico creek can act as holding lies for fish moving through faster water or, in this case, water that has a high sun overhead and predators likely nearby. PHOTO BY NICHOLAS CONKLIN

While the holding lies are not as deep and protected as the pools just covered, these areas do act to hold fish and can be productive holdover locations. However, the most common mistake here is the amount of time spent fishing these sections. Too often anglers "overfish" in certain areas while underfishing other lies and not expanding beyond a singular technique. This pocketwater often requires a more precise approach.

In the hierarchy of zones of safety for holding fish, the pocketwater or boulder field is not as protected in most rivers. The zone of safety is more compact, and fish will more often than not be in smaller groups. However, that does not mean it is not worthy of legitimate fishing time. Due to the flow of current around the obstructions, these pockets can be highways for bug activity and, in most situations, can help focus fish due to the funneling nature of this water.

Effective anglers take a downstream-to-upstream approach. Again, they start close to the bank before working their cast out, often described as fishing inside to outside. Here an angler fishing a dry-dropper begins their approach only feet off the bank before reaching out to the water flowing from left to right behind the obstruction.

What's important to consider here is fly placement behind the obstruction. Anglers should key on the moving water directly to the left or right of the boulder, as well as the slack water that spreads around and downstream.

The softer, steady water that emerges anywhere from one foot to several feet behind the boulder affords a great path of least resistance for trout. A common mistake anglers make is not working the steady-paced water behind the boulder first. They tend to gravitate to the fast water on either side of the boulder, where poor fly placement can lead to snags or the flies getting caught in the circulating, eddying water directly behind the rock. This water tends to be the "dead zone" in this regime; flies will

With a tight approach, an open-minded angler can pick off some fish hanging near the bank. A single floating fly here allowed the angler to work upriver, fishing shallow spots without fear of snagging and losing flies.
PHOTO BY NICHOLAS CONKLIN

get caught in varied current speeds, and any attempt at a natural drift will be lost.

It may be beneficial to attack it with a varied technique cadence given the time of day or year or water fluctuations. Starting with a high-stick or nymph approach, followed by a surface and/or dry-dropper scenario, and then a streamer approach, affords the angler every chance at success.

Success can still be found when an angler works through this section rigorously with quick and efficient fly placement. As discussed in previous chapters, this is a fitting example of a water regime that can be fished on a timer.

By mentally blocking off a 20- to 30-yard section of water and only committing a set amount of time, anglers become more conscious of their time and focus on a few quick, sunken-fly approaches. Depending on the water, anglers should consider shifting to smaller and lighter flies.

A dry-fly angler should attack pocketwater with vigor due to the varied currents and the opportunity to fish above rocks and the slack water to each side that can hold fish. It's wise to keep floating indicators and other large bobber-like items away in this water. There is too much commotion and not enough time to drop flies into the proper zone here. The size and water resistance of these items act to keep flies from sinking into the strike zone.

On bright, sunny days or during periods of low water, the effective angler needs to work through these sections more quickly and with an eye on the type of shadow they are projecting on the water. Natural obstructions can be beneficial when immersed in these bright, obvious scenarios. Thick brush, trees, and even making a few initial casts from a crouched or kneeling angle can help keep the presence of your body and casting motion less apparent.

Timing here is important, and it is beneficial to make mental notes of which type of water should be re-targeted during the return to the vehicle. Pocketwater is a notable example of quick fishing opportunities that will change over the course of a day. The approach can be simple and streamlined. Just because the water sat empty at 9:00 a.m. does not mean fish have not moved in later in the day.

RIFFLES

The slower-moving, very shallow sections of river known as riffles can be intimidating to wading anglers. While these are typically the pass-through or pull-over-and-piss spots for boaters, wading anglers can effectively dissect these areas and pull fish from some unlikely locations.

Due to the nature of the current speed and the amount of sunlight that is often afforded to these areas, substantial amounts of bug activity can occur. In these less pressured stretches, insects can safely hold on to the smaller rocks and boulders. While ideal for insect life, these waters are the most under-approached targets for anglers.

Why? Because of the current and lack of recognizable holding lies for trout. This causes the less ambitious angler to look up- or downstream for water with more noticeable lies. However, this drive-by water should be fished.

While it may take more time to notice any depressions, slicks, or seams, these areas can hold trout. If anything, these fish are less hassled than those in the more written and photographed waters.

Due to the depth and physical nature of riffles, the cautious angler will benefit from approaching slowly and subtly. This water can yield positive results when blind-casting surface and subsurface flies. With an upstream, quartering approach, anglers can cover wider sections of the river to increase the odds of a take. Two other effective approaches often missed by the angling public are the swung fly and the high-stick or Euro approach.

These tend to be productive areas for the angler who wants to swing very lightly weighted flies such as soft hackles. Whether on a single-handed or two-handed rod, this fly setup (and often multi soft hackles) can cover water while keeping the fly in an effective short-depth zone.

Due to the lack of aggressive currents or obstructions that would create multiple flows, riffles can afford anglers a more consistent drift and less time focusing on line management. While those management skills are critical, this can be a good portion of water to warm up on and revive your casting stroke and line management while covering fishy water. What will set the successful angler apart on

With two soft hackles and standing in riffle water, the angler can swing both sides of the river, looking for pockets or small depressions where a fish could hold. PHOTO BY NICHOLAS CONKLIN

Regardless of the type of water regime or size, anglers should always consider the path of least resistance. What is the most efficient place for a fish to hold, to conserve energy while still maintaining the ability to acquire calories?
PHOTO BY OLIVER SUTRO

this water will be their ability to cover space and spot small depressions and holding lies. Covering space, especially when blind-casting or making down-and-across casts, and effectively placing the line consistently across the river is necessary. The focus here should be on accuracy and thoroughly working the water.

The spotting portion of this process may take some time to develop but will be a key piece in better understanding and evaluating other river sections. Western rivers like the Madison offer many of these long, steady rock fields. While initially intimidating, anglers should always look for the smaller indentations in the river bottom. They may be tough to decipher, but these small depressions will hold fish throughout the day. This is where an angler can set themself apart on larger, heavily fished rivers. Instead of focusing solely on an angled approach to the bank and obvious obstacles, they look to small indentations and subtle changes in depth more toward the middle portion of the river. These can be safe havens for fish that have been battered and beaten by anglers while holding along sections closer to the bank. Whether disturbed by anglers walking the bank or boats floating by, these fish will sometimes scoot to these unconventional lies to avoid harassment or threat of predation.

One straightforward way to do this before stepping into the river is to simply observe. By standing back, and ideally walking both sides of the bank, and watching the riffle portion during various times of the day (think sunlight and shadow changes), the willing and patient angler will begin to detect fish movement. While the holding positions will be subtle, they exist, and many fish will lounge in these holdover spots throughout the day.

Floating anglers have the upper hand in this scenario, as they are more apt to spot and quickly adapt to a possible holding fish. But a few quick casts in the middle section of water can lead to a hooked fish that has heard and seen several boats drifting within yards of the bank. One thing to always remember is the fish can be pretty much anywhere, so when in doubt, make a few casts.

Technique-Based Casting

In this chapter, we discussed how anglers can become better prepared and more adaptable by instituting some casting principles. Often talked about but forgotten when an angler steps into a river are the technique-based casting abilities of the angler. To outdistance, or "outfish," another angler, they need to arrive at the water with a solid foundation and develop skills to correct flaws throughout the course of the day. While the end game is not always to outproduce or stack numbers like a professional bass fisher flopping four-pounders onto the deck, technique-focused casting is a critical step in overall enjoyment. And that increase in enjoyment can be accomplished by more than catch rates.

Casting issues and failures cause frustration, more so than not hooking and landing fish. While most of us can recall how we learned to cast, we may not remember when we transitioned to learning a specific fly-style based cast. Whether by watching old TV shows like *Walker's Cay Chronicles* or *The Saltwater Sportsman* or by poring over the pages of *Fly Fisherman* magazine, we have some compelling memories. Unfortunately, those memories seem to slip away once the actual fishing begins.

After covering and reiterating the principles of the effective cast, we, as hopefully more effective anglers, should now take the time to observe and dissect the casting of other anglers on the water. While we examine how they approach different technique-based casts, we should be able to dissect the parts that we need to apply internally. While you may feel like a wild-haired, snoopy neighbor, peering with bloodshot eyes through a slit in the curtain, we can't emphasize enough the ability to watch, analyze, and learn from other anglers.

THE EURO CAST

The Euro cast is one of those tricky areas, where most of your thought process and understanding of how to generate the right amount of mass (fly line) outside of the rod tip to effectively load goes out the window. Due to the lack of a tapered fly line and therefore no head on the line, the motion to deliver flies is different from your traditional cast.

A rounded or oval-style rod path will aid in loading and quickly depositing flies. For anglers switching between traditional fly lines and mono-style Euro rigs, the arm motion and stroke will need to change. It becomes less about unrolling a fly line and more about the hard stop and placement of the flies, due to lack of mass because of the nontraditional fly line. PHOTO BY OLIVER SUTRO

Those who are fans of inconsistent application of power, or too much speed to the stop, rejoice! You've found your cast. To keep the observation points clear, we will cover a common casting fault and a correction here.

The angler makes a more elongated stroke, attempting to "load" the rod in the same motion and cadence they would while making a cast with a more traditional weight-forward fly line. They fail to generate load, and the set of flies moves in an arc and does not land authoritatively in the water. This approach creates two problems in efficiency.

First, the angler will, throughout a fishing session, put more power and energy into the cast while not achieving the desired result. This will inevitably lead to a tired and frustrated angler. Second, the multi-fly rig will move and land inconsistently, not allowing for precise fly placement.

The final issue is fly depth. Without an authoritative stop, flies will land softer and typically take longer to reach their effective depth.

Correction? The plane or path of the rod tip must change.

Remember, there is no weighty fly line to load the rod and unroll your flies. Here, the successful angler will use a much shorter, direct-plane approach. This oval cast with a high stopping point not only reduces stroke length but also allows the angler to keep the rod and reel close to the body, allowing for more efficient energy use. This aspect will be key as the day wears on to prevent fatigue.

The last point of correction comes down to the ability to get the flies to land quickly and sink into the strike zone. While water does slow the sink rate of flies, what anglers should keep in mind here is the speed at which they land and begin to sink.

A softly thrown rock, with an arching motion, will sink at a different rate than a rock of the same size thrown at a greater speed at the water. The ability to break surface tension is key.

Flies "lobbed" into the water will take more time to break that tension, and you will notice the sighter does not become taut until well into the drift. The angler will achieve an ideal fly depth when they can effectively land their flies quickly, allowing them to drop into an active feeding zone. The sooner this happens, the longer more sustained drift will occur.

THE STREAMER CAST

Hindrances to casting large, heavy flies can come from a variety of root causes. One of the most common is simply the lack of time most anglers spend casting them.

As anglers move between techniques, they need to be mindful of the weight and type of fly they are trying to deliver. An angler cannot take the same tempo, stroke length, and rod tip path they took in the early morning with a light line and size 20 Trico and replicate it with their 9-foot, 8-weight streamer rig, with a 240-grain line and a multi-articulated cone-head streamer. However, one thing anglers should always be on the lookout for is doing too much and casting too far when changing to this style of fishing.

The temptation is there. A big powerful rod needs a big powerful cast. Too much line gets stripped out, and the angler ends up spending more time trying to haul line and manage a pile of running line than fishing. The first area for the interested voyeur to watch is the loop during the cast. Whether it is on a full-floating or full-sinking line or sink tip, the generation and formation of the loop need to change to account for the added weight and its distribution along the system. Author, fly tier, and innovator Bob Clouser has spoken at length about how the weighted fly and/or unweighted fly angler needs to adjust their plane of thinking when it comes to effectively delivering flies in this scenario.

Known the world over for his classic Clouser patterns, Bob's lead-eye flies were designed to suit his specific fishing needs in Pennsylvania and other areas where a sparsely tied but fast-sinking fly is necessary. Soon after gaining acclaim for his pattern, he realized that clients and anglers were struggling to deliver the fly. The traditional over-the-shoulder, close-to-the-head cast wasn't going to cut it.

As Bob jokes, "It only took a few flies to the back of the head for me to figure out the cast needs to change." So, with input from industry friends and fellow anglers, he began to develop a rod tip path that keeps the fly away from the body while also allowing for more loop control and use of energy.

What resulted was what Bob calls the "inverted loop." This allows for proper rod load while keeping the denser line and fly moving in a safe path. It is a great cast to develop while fishing from an elevated platform such as a drift boat.

casting without a strike, fishing is a challenge, and on many of the highly pressured western rivers, the cast-to-hookup ratio is becoming smaller. Breaking it down into a few phases, from active fishing to hook sets, rod angles, and landing strategies, you can easily outpace the person who just waded the same water.

Remember back in grade school, standing out on a baseball field, trying to not fall asleep? While a teammate in right field sat building a mini fort from grass and the second baseman had his finger in his nose, as someone with cupped hands yelled from the dugout, "Get in game position"? Ideally, the move to fly fishing is much more impassioned than that of a third-grade, rec-league T-ball team. Anglers in active, focused positions stand a far better chance of responding quickly to a strike.

When deciphering whether an angler is actively fishing, it's not always as simple as seeing who has their eyes forward and is managing a focused drift. They may be fussing with a fly box, scratching their ass, or otherwise acting like Tommy, the lackadaisical outfielder. But when the cast is made and the fly hits the water, it's game time.

Observe the position of the rod and the angle it makes following the cast. Is the rod low and the tip not tracking or following along with the presentation?

This is a common slip of the swing angler. As they cast and mend and let the fly swing into the zone, they tend to drop their arms. This added distance and lack of contact can affect a hook set in two ways. One, more of the line (and, in a shooting head system, the running line) is on the water, creating drag and thus deadening the response to a grab. Also, with a dropped arm and hand position, the response time is slowed. While on the swung fly this is not always bad, what tends to happen over time, and time without contact, is a more aggressive motion is made to "set the hook." This fault is common among anglers stripping streamers and other larger sunken fly techniques.

For anglers floating an indicator, inactive periods can be recognized by the angler making a

The experienced angler focuses on small strips of line so that she can quickly respond to a fish. Fishing "shorter," with an emphasis on line management, will lead to a higher hookup rate. PHOTO BY CAPT. COLLINS D. ILLICH

Large, heavy flies require a different casting stroke than nymphs or dry flies. Anglers should look at the inverted loop or pendulum-style cast to keep a heavier line and fly away from the body. PHOTO BY NICHOLAS CONKLIN

The wading angler can also benefit from initially inverting the loop. However, one alteration is that while the back cast is inverted and remains below the tip of the rod, the path changes on the forward stroke. This allows for the heavy line to cut beneath the wind while keeping it a safe distance from the angler's body. The more exaggerated open loop on the forward cast keeps the fly from colliding with the line and leader. This can also be helpful when casting cumbersome multi-streamer rigs. While the inverted loop and pendulum-style cast may change in angle, the effective caster keeps the path of the rod tip consistent.

A laser-tight shot through a screen door loop is not always the goal with streamer fishing. The parts of the equation require a different approach. Your guide will thank you if you can keep another cast in your back pocket and keep from buzzing their head all day.

The second portion of the streamer cast evaluation comes down to distance, or, rather, the perceived value of distance.

While applicable to both the floating and the wading angler, the idea of stripping off line to cover water tends to distract more people than it is worth. Instead, anglers should watch those who make a shorter, more repetitive cast. When an angler can manage their line, by keeping it off the deck of the boat or from catching the surface current and moving downstream, it means more time with the fly in the water and more chances for success.

Being an Active Angler

Now that the various casts are perfected and you have spent considerable time evaluating (judging) and making serious strides to perform a better cast, the intrepid angler needs to evaluate how other anglers fight and land fish.

Anyone who has spent serious time chasing anadromous, catadromous, or otherwise named migratory fish has had this experience. You've spent hours, days, or even in the direst situations seasons, casting and swinging flies. After countless casts and steps, you feel an ardent tug, only to suddenly forget everything you've been taught, lifting the rod tip and setting on a fish like it's a game of tug-of-war.

While the odds are most of your freshwater trout scenarios won't require countless hours and days of

The angler is in a ready position, with contact with his line hand and body in position. He thoughtfully manages his line, preventing as much slack as possible. At the end of the drift, or if a fish strikes, he will be ready.
PHOTO BY NICHOLAS CONKLIN

half-hearted mend and looking away. We will reference this action from the point of view of the floating angler, but it is equally common for the knee-deep person, as this technique seems to allow for down periods. This commonly happens as someone is merrily drifting down a seam and hasn't had an indicator twitch in some time. They get told to make a cast, perform a subtle upstream mend, and watch their floater. As the mend occurs, nine times out of ten, the angler drops their rod and briefly looks away.

I've seen it as often as I've seen my bank account mysteriously diminish. I am never genuinely surprised when it happens, and yet it occurs more often than it should. This glance away puts the angler at a disadvantage, as more often than not they will not notice indicator movement.

If the guide yelling "SET!" wasn't enough to shock people back into focus, the true number of missed fish should. As the line piles and the rod tip drops, the reaction time and body movements of the angler almost need to double to catch up to the slack created. Arms are now forced to make a longer, more aggressive path to recover line, and by the time it is realized, there will be nothing on the other end.

One way to combat this is to fish a shorter-than-normal line as the day wears on. While not seeking to dispel the hardened wisdom of the almighty river guide, there is a science to adapting this approach, which can't be disputed over a half-burned Marlboro. No matter how many cups of gas station coffee and Highway 287 truck stop uppers ingested, anglers will tire, and these slack periods will increase. While the adrenaline is pumping early, it will burn off. Anglers should consider this the time to shorten their manageable line and become hyper-focused.

Less slack means less chance for the river and the fish to pass by your flies. By maintaining more direct contact, you will increase hookup rates and improve speed and accuracy when repositioning a rig. Less line on the water will act to combat the potential for slack and the lack of direct contact when setting the hook.

While a number of experienced floaters faced frustration with the increased crowds and lack of etiquette, it reminded many that simple, friendly communication with those new on the water helped everyone have a better day. PHOTO BY CAPT. COLLINS D. ILLICH

CHAPTER 5

FLOATING STRATEGIES

During the boat and raft buying boom of the early 2020s, thousands of river and lake fishing vessels were purchased. While professionals and working types made strategic maneuvers to profit from the lust and limited inventory, many new rowers hit the rivers. With license sales skyrocketing over the same period, the boat buying frenzy provided a compelling way for anglers to beef up their fishing résumés.

But not all anglers moving from occasional guided trips or wade trips possessed the proper foundation of rowing and drift boat fishing skills. With varying degrees of experience and willingness to learn, many spots in the West became congested due to this new and often unaware crowd. Like a high school driver's ed class, some prepared rowers floated by with ease, while the unprepared floundered and clogged the water.

An angler's ability to sit and watch is a crucial skill that sadly falls under the easier-said-than-done category. Studying the fiberglass flotilla, rafts, and other roto-molded, space-age angler transportation devices and knowing why they move the way they do is vital to the floating angler.

You'll often notice that the paths of hired boats follow some standard lines. That makes sense, right? They are on the water, know the fish holding and safe navigational lies, and do all they can to put their clients on as many fish as possible. If you spend enough time near a stretch of river, you can predict this boat "hatch" and dissect the adventurous anglers versus those following the leader.

Some of our less-than-scientific research on traffic levels and peak times comes from watching boat ramps and standing on western river bridges observing. Just as a dry-fly angler will stand at attention to see which color variant of bug they should tie on, boating anglers should do the same when scouting water. This stage shouldn't be about watching bugs or looking at anything other than the path these boats or wading anglers are taking. The likely reason paths along the river are beaten down and easily accessible is that they lead to good lies.

If you have the luxury of fishing the same water multiple times a year, it may be beneficial to watch how different boats float and fish in different portions of the river given the time of year, time of day, and weather conditions. Certain angling paths can change from hour to hour depending on weather conditions. The most common difference is the transition from sun to clouds. This may not be the most obvious change of course for the floating angler. However, if you can get up above the boat, from a bridge or nearby road, a quick photo of boats floating by can help provide a daily or monthly reminder of the different routes that boats will take.

As critical as it can be to note the most common route boats take, it can be equally important to remember where boats don't go. Outside of why they won't float there, there may be reasons they don't want to fish these lanes. Angler and rower aptitude, current flows, and obstructions all influence a boat's path. Is it merely a safe water-flow situation? Are there seams or sections that should be explored but, for whatever reason (angler casting skill or float time), often get overlooked?

Outside of the formidable expense and maintenance required, boats are more than just a badge of honor. These floaty things can allow you to explore different sections of water, and floating just a yard or two of river left or right could yield results. Why not try the path down the middle?

During a recent fall float trip, we experienced firsthand when it is wise to follow a fellow boat's path. Due to extremely low water flows, working down the river presented many boulder fields and long, shallow paths. In an attempt to not crowd the lead boat, the tailing boat made more noise, dragging the bottom. While trying to provide space and limit noise so as to not interrupt the other boat, more ruckus was caused and more fretful internal conversations were had. Often it can just be easier

Credit ratings were just numbers when people were driven from their offices and workplaces and descended on boat and raft makers. This situation resulted in record sales during the COVID pandemic. This boom resulted in many inexperienced rowers on the water. PHOTO BY OLIVER SUTRO

and more pleasant to follow behind a pro. On meeting the guide boat at the ramp, a conversation was had, and the result was that the guide was trying to get our boat to follow his downriver. He was familiar with this stretch and wanted to help.

Bottom line: There are two ways to approach following the path—trying new water when fishing or following along in what you have done or seen done before. Sometimes following the flow of traffic is the best way down the river.

One important consideration before the trip is clearly understanding potential threats. These threats may not always be as obvious as a downed tree or massive boulder in the middle of the river. These danger zones can lurk just beneath the surface, waiting for their barred teeth to grab the bottom of your boat and jostle everyone around. Clear communication is key between the angler in the front of the boat and the rower. The job of the rower is not only to get the angler in the perfect position to make a cast but also to ensure a safe float.

In the book *Float-Fishing Strategies*, author Neale Streeks breaks down maneuvering upcoming obstacles. No different from a checklist for packing, Streeks suggests rowers develop a mental list of boat movements. He stresses recognizing potential dangers early and beginning evasive techniques as soon as possible. The rower must begin to build momentum with their oar strokes to position the boat to avoid the magnetic attraction of the current to the obstacle.

That book, along with rowing-oriented courses, is truly the best way to learn a stretch of water. While guides may not openly enjoy getting the third degree from a client about a section of the river, they do value a conscientious angler wishing to learn how to row a river safely and courteously. We do suggest guests clearly state what their learning objectives are, which ideally are not to steal the guide's hard-learned, income-generating water. These steps should be completed well before the purchase of a raft or boat.

The wise rower not only focuses on the angler's position but also keeps their eyes up, watching the movement of other boaters. It's important to be constantly moving your head to ensure both an effective casting position for the angler and overall boat safety. PHOTO BY CAPT. COLLINS D. ILLICH

As the boat approaches a large boulder, the rower can build momentum and begin the backstroke needed to point the bow at the obstruction. PHOTO BY CAPT. COLLINS D. ILLICH

The many nuances that go into the design and building of a river boat are suited to certain situations and preferences of the rower. While paint styles and cup holder allowances may seem pertinent when browsing online, what matters is the feel when sitting in the middle seat. The height of the gunnels, the length of the boat, and even the leg locks all vary, and all deserve a test run before a deposit is placed.

The outfitting stage should be of the utmost importance when it comes to the style of craft best suited for the water you will be on the most. With floating devices ranging from small prams to large motor-capable boats, a serious dive into necessities is required. Not all drift boats are best suited for the smaller, more protected waters of Colorado. A raft-style boat may only act as a sail on larger, wind-blown Montana rivers. Buyers need to consider how much time they will spend on specific waters and the location of these waters before deciding on the shape of the vessel. Will your preferred river have the correct amount of launch access for a traditional drifter, or is a lightweight, portable raft more suitable?

The investment in fishing with and picking the brains of experienced local guides will drastically aid in answering these questions. During these trips, eager rowers should also take the time to evaluate other boaters and truly question why they do what they do. These initial conversations can further the learning curve when it comes to boat ramp etiquette.

Unfortunately, one of the major points of contention between the local "working class" and seasonal anglers is how they behave at the launch. Due to the number of new boaters, the often unwritten rules of ramp and river etiquette have caused heated exchanges. Tired, overworked, and frustrated guides may not always have a helpful, willing attitude toward a recalcitrant rower.

The best course of action when overtaking a float boat either at the ramp or mid-river is caution. While distance from other angling parties will generally vary depending on the river, keeping a safe space from others will always be appreciated. Like a coiled snake, the same approach should be

With the anchor still deployed, boaters can more quickly load and trailer their boat off the ramp. With the anchor down, boats won't move and get pushed downriver and off track while loading. PHOTO BY OLIVER SUTRO

taken when boats are anchored waiting for their trailers. Pulling over and anchoring 50 to 75 yards up from the ramp will give others the time to load and pull out. Cramming into a confined space and being slow on the trailer is not a way to make friends in the West.

The same sense of urgency should apply when the boat is on the hook. Anglers should have promptly departed the boat to allow it to be quickly hooked, cranked in, and pulled from the launch. This is not the place to stand around and take up space. Post-trip conferences and the removal of waders should be done in the lot, away from traffic. While this may be obvious, anglers can develop tunnel vision after a long day on the water. Whether overcome with happiness after many landed fish or hunkered down due to rain and wind, lingering where others need to maneuver occurs all too frequently.

As referenced in chapter 2, having a streamlined, organized boat that is ready to launch will prevent drawing scowls from those in line. Efficiency will come from organization and practice launching the boat and backing away a trailer. We can't stress enough that new or seasonal rowers should develop a strategy for loading and launching, with practice. This practice time should come in the off hours when ramps are empty and others' fishing time is not being encroached on. Each trailer responds differently in different conditions, as do concrete ramps. These differences can be learned by both watching others launch and practicing on one's own.

All safety measures regarding flotation devices as required by fish and wildlife departments should be adhered to. What likely doesn't get covered is how to respond should the boat be pinned or capsized. While seemingly garish and uncomfortable during casting, life jackets can be vital and required by law on many rivers.

The teamwork component cannot be stressed enough. This is especially true for new and intermediate boat owners. While the thought of rock spotting may never have occurred on countless guide-driven trips, the sticks are now in your control, as are the lives of those on board.

Rowers on new or unfamiliar water also have to understand that they will hit rocks and drag the bottom. It is part of floating in the West. The key is understanding what will just scrape up that fancy new boat and what could have a serious safety impact on the occupants.

Casting before the Float

If you're not familiar with casting an indicator, multi-fly rig, or streamer in a drift boat or raft, practice is the only way to improve. This book will document some steps to prepare for that float trip and the casts and styles needed to achieve greater success.

To help mitigate the creeping gray hairs along the hat lines of western fly-fishing guides, we can't emphasize enough the value of casting before a trip. Whether it's in a guided boat or a friend's vessel, some practice time will be another step to further separate you from other anglers. If only to keep your fly in the water longer, some prep casting is vital. The distinction here lies in the growth from casting lessons to casting to fish in live fishing scenarios. Once the classroom casting steps are built, anglers should then move on to more challenging real-life scenarios.

Even if you don't currently own a boat, many household and neighborhood items can offer a "challenging" casting platform to help provide a realistic point of view. A stool, ladder, chair, or bench can provide some of the necessary foot and body placement to replicate being in the bow of a boat. This elevated platform can help with the angle of the casts as well as build awareness of body placement on a moving object.

A simple, inexpensive section of leader, an indicator, and a tuft of yarn is a fine starting point. Without weight or flies and just a bobber or yarn to replicate some mass on the line, casters can reorient themselves before a day floating.

The angler looking to fully maximize on-the-water time will spend this time perfecting casting in all possible scenarios. The ability to develop a cast with both hands, off of both shoulders, cack-handed, side-armed, high overhead, or windshield wiper style will afford more time fishing. Trees, bridges, and even the other anglers in the boat are all obstacles that can be prepared for in some way or another before getting on the water. Halfway through a float trip is not the time to discover that you should have practiced casting backhanded.

Wind will also play a role in any western fishery. It may not seem like it's blowing when you hop out of the truck and are standing around the boat ramp, but odds are you will face it on the river, and you need to be prepared to adjust your angles. Waiting

Even in ideal floating conditions, anglers should be more than comfortable casting off non-dominant shoulders and away from other anglers to get the fly into fishable water. Waiting until the day of the float will ensure much of the day is frittered away "practicing" instead of fishing.
PHOTO BY CAPT. COLLINS D. ILLICH

The authors were only yards from the ramp when they began casting. By arriving prepared, with flies rigged, they were fishing well before the bridge, where most boats were just beginning to get lines wet and working out delivery kinks.
PHOTO BY OLIVER SUTRO

until you are 200 yards downriver to start practicing your cast, avoid hitting the rower, and attempt some decent drifts is too late. Your fellow angler, rower, or guide shouldn't be the one to help you once the day has begun. Yes, the nice ones may offer guidance, but anglers who truly wish to out-achieve others on the river will commit to pre-fishing practice.

Confidence in casting ability will help overcome any perception of nerves or uncomfortableness. Whether with familiar anglers or a new guide, energy and new environments can lead to nerves. The best way to counter this is just time on the water, around other anglers, and in semi-pressured scenarios.

Now, we all should know of the famous author John Gierach and his musings on fly fishing being a "bohemian business that sets you outside of the mainstream of society." All accurate and likely why many of us feel the need to side-step daily responsibilities and job duties to abscond into the woods wearing neoprene and hats that would draw stares from even the most granola-munching, wannabe Haight-Ashbury hippie. But there is value in being able to execute casts and proper drifts with the eyes of other anglers on you. The paramount idea in this text is to watch and learn from other anglers, so the awareness of others watching you should not cause poor form.

A local park can be a suitable place to practice techniques, as many non-angling eyes will be focused on you. Block out the gasps and finger-pointing from snot-nosed kids and horrified nannies. Don't worry; when they tell their parents later on about the "strange person waving a long wand in the park," they won't be believed, and they will be hastily tucked into bed for the night.

What this can do is provide some faux pressure to help you better develop focus techniques. Similar practice needs to be applied if you intend to move between an indicator-type system, a dry-fly application, or heavier, larger flies on tips and lines of various densities. That evening, through a screen door loop you have developed on your dry-fly cast, it no longer will work when throwing larger flies and sink tips or full-sinking lines. You'll be best served to practice opening the loop, inverting the cast, and developing a level of comfort to quickly move between multiple styles. The ability to accomplish this will easily set you apart from most other anglers.

Preparation prevents failures; more critically, proper preparation is the key to success. Practicing bad techniques only embeds bad habits, which leads to frustrated fishing. We can't recall how many thousands of times we've been told by a crusty old angler, on the phone, in a hastily one-finger-typed email, or on the floor of a trade show, that they know better based on age.

"Well, I have been doing it for forty years!"

"You may have been doing it a long time, but that doesn't mean you've been doing it correctly."

Time is not an indicator of skill. Age and/or gender does not magically bestow talent. Sorry, old dudes, you can't sit on your butts all year, never take a proper casting lesson, and somehow be handed skill. Equally important, you can't walk up to a rod rack and buy the skills needed to be a successful angler and caster. It takes time and well-focused effort to learn the skills to be an effective angler.

If you're not practicing, practicing right, and pushing your comfort level, average angling and casting will be the result. Which is fine; it's your leisure time and leisure money. Like the choice of beverage, people should be free to attend to their off hours how they want, without pestering. Professional lessons and training opportunities are one component to improving this practice skill. These skills can differ whether floating or wading, and, as we discuss in future chapters, understanding how both work together can build a more well-rounded angler.

Time and Distance to Active Angling

As we've discussed, lack of practice and preparation leads to a failure of time and distance to active angling. Whether you are a private boat or a hired

Angling pressure at popular western spots does not always mean a lack of willing fish close to the launch. While it's important to row away from a crowded ramp, floating anglers should have rods rigged and line stripped out as soon as the rower gives the go-ahead. PHOTO BY NICHOLAS CONKLIN

stick, the distance from launch to a line in the water averages about 75 to 100 yards down the river. We can posit all of the reasons why a hired guide needs some time to get their clients ready, like cleaning up the bags and other shit they dragged on board and then getting rods in hands and a few dump casts out. While important, they, unfortunately, miss some often-untouched water.

Even experienced anglers can be at fault for not being ready to fish.

On a stretch of a western Colorado river, some of the best streamer shots, early in the morning, are within 50 yards of the boat ramp. Once, while still wiping the fog of the previous night's adventures from my eyes, I (Nicholas) missed over a half-dozen fish. While haphazardly slapping the fly down, I was not in game-ready, active position. I was not expecting fish to be so willing so close to the heavily trafficked park. Despite having the right fly, decent placement, and the constant advice from the guide to "get ready," I missed every strike.

So even with all of the other components well practiced and in place, I still was not ready to fish.

Like everything, preparation in scenarios such as this will distance the occasionally successful from the consistently effective. Also, do some critical independent thinking when looking at the river.

If you have floated this stretch before and have some capability, it'll impress your guide to reach out before the trip and mention it. Ask them what else you can do to prepare before getting on the boat. What do they see from anglers who are productive? Is this a typical pattern time of year and stretch of water? If bringing a personal rod, maybe mention how you want to get "active" fishing early and that you have done the work, developed the skills, and are ready to fish.

The thought in most fly fishing often applies here, and it essentially boils down to just having a fly in the water. It doesn't have to be the perfect size or the best shade of thread wrap or even the most realistic pattern. What it does need to be is in the water and in the correct zone. The "in the water, in the game" idea can help anglers pick up a few fish. Those looking to further improve will be mentally prepared and ready to fish. That, combined with the physical preparedness of tackle, will save them from experiencing the missteps this author did.

If floating privately, this space and time can be used to explore untouched water. Whether starting a float or ending it, usually there is a good stretch of river that doesn't get appropriately fished. The day is about over, you've been fishing for several hours, and most of the last 100 yards before getting off the water is spent hiding empty beer cans and putting rods and flies away. Regardless of the success of the day, most rowers will move through the float quickly as anglers sit down and arrange gear. The floating angler should always consider this water active and perform casts even up to the stopping point.

The same goes for those just starting a float. This right-off-the-trailer time is too often spent fussing with gear and getting leaders and flies unhooked. While it seems the portion of water with the most traffic, consider the lack of pressure exerted under the surface. A simple glance while standing on any bridge near a ramp can highlight how most hired and even private boats tend to get 50 yards or more downriver before actively fishing. Considering that most anglers do not approach the next 50 yards in the active position, the opportunities close to the ramp should be obvious.

The mental roadblock that you need to get downriver and "away" from the most obvious sections limits success. Like the wading angler who always feels the need to hike miles upriver to escape the crowds, the fish don't always see it that way. The proximity to the launch should not hinder a floating angler's approach. While not wise to anchor or back row in these congested areas, anglers should always be looking at this well-trafficked but forgotten water.

Fishing Strategies

Fishing strategies out of the confines of any boat or raft will require different spatial awareness. In chapter 8, we cover the techniques and tackle for anglers looking to improve their success rates while fishing from a boat. In this section, we want the thoughts of the rower to be on not rod length or mending styles but the mental approach to boat operation. While a paced and thought-out game plan before the boat is slid off the trailer is important, so is what happens a few yards downriver.

This is an area where the rush of excitement and the need to multitask may become overwhelming,

Rowing can be an exhausting mental and physical exercise. When frustration mounts, often from scraping bottom or banging rocks, a fishing team should pull over and take a deep breath. A wise Montana fishing guide often reminds us that yes, you will hit rocks and bang the bottom. That's just how fishing in the West goes. PHOTO BY NICHOLAS CONKLIN

even for those who have taken rowing classes, floated with guides, and prepared for the water they intend to fish. One wise piece of advice we've found is that when things are moving too quickly, anglers are struggling to catch fish, or obstructions are being hit, just pull over. By rowing to shore and taking a few minutes to catch a breath and think, rowers can essentially hit the "reset" button on the last river miles.

This is time off the sticks, which can help to calm nerves, slow the heart rate, and further reduce wasted energy. During this break time, tackle can be rerigged and preplanned tactics and goals can be redefined. This is also a time for the rower and angler to further discuss what they saw, obstacles missed, and how to better approach the river ahead.

Like all statements in this book, we aim to remind anglers that efficiency of movement can result in better use of energy. The ability to stop and correct mistakes and energy burn can be key to long-term success.

While not always the easiest to locate, quality public-access wading water can be found all over the West. An unfortunate mental hurdle most anglers need to leap over is that fly fishing can be done only from a boat or raft. PHOTO BY NICHOLAS CONKLIN

CHAPTER 6

THE WADING ANGLER

While the previous chapter focused on techniques tailored to the approach of the floating angler, it is equally important to examine and sharpen wading skills. Even if you plan to fish primarily from rubber or fiberglass, the angler looking to really improve will seek to grow and adopt successful behaviors in both disciplines.

Improving wading skills will afford anglers a more complete toolbox, whether they spend more time on foot or in a boat. The key is the angles and perspectives. Each system can provide a varied view of the water, and it is vital to always approach water from both perspectives.

We aim to further the idea that wading access in the West is not only achievable but also viable to a larger population of fly anglers. Because it is a more feasible avenue, honing skills is even more paramount to out-succeed fellow anglers.

There are countless affordable and reasonable access opportunities spread across the Northern and Southern Rockies. While it may seem über-trendy to assume every inch of water in the West has been scooped up by the ultra-wealthy, there are still places to fish. Not all boat ramps are packed with Teslas sitting on a charge, and you can still find access points that don't threaten invasive legal action and public flogging.

You don't need a boat to access quality fish. Truthfully, we spend nearly 70 percent of our angling time each year on foot. And honestly, it's a far more enjoyable and connected way to fish. Quality fish and fishing experiences can still be found in substantial numbers across the western United States.

As of 2022, Montana alone listed over 330 public-access fishing sites. In conjunction with Trout Unlimited, they have generated a comprehensive map to highlight these areas and educate anglers. Similar maps and information can be found in each western state.

Anglers and conservationists are often bombarded by the media with topics of overreaching private landowner control and limited access, but much of the West still holds accessible water. While it certainly is not a regenerating resource and access has been limited, anglers should seek out these lesser-known public waters as more pressure is put on the overall system.

There are thousands of assets to help the wading angler prepare, including options like Google Earth and digital river maps. By simply driving along the river, looking around, and buying a few beers in a small river-rat bar, adventurous anglers can dig into the local scene without committing their life to the search. Since the advent of the internet, the wealth of knowledge and, alternately, the fetid, leaky pipe of garbage information have often battled it out on our screens. While we now have access to a lot of detailed information, we are also often confronted with negative and exclusionary attitudes.

We feel this is a great connection to the overall message in the book. Like bemoaning crowded and heavily fished water, the same can be applied here to digital resources and research opportunities. The ability to digest quality information and carve out what is just prattle from bored anglers sitting at the computer versus what is valuable in both positivity of message and technical information is difficult. While forums, message boards, and other aggregate-style sites still have their place, we prefer a combination of local fly shop intel and the use of access site apps.

Using and Studying All of the Resources for Current Information

Whether in person or through individual fly shop websites, this type of hyper-focused regional, seasonal, and condition-based information still reigns supreme. Evolving past the initial gripe of big chain stores "stealing" the business of local shops, the real

Active anglers will use printed, or hard-copy, information and actively compare it to what can be found from phone apps and modern technology. Changes in weather conditions and water flows can be more easily monitored. The wise angler will do more than just rely on past float trips and perceived knowledge. PHOTO BY OLIVER SUTRO

purpose of the local fly shop emerges. This is the opportunity to truly refine a skill set and learn about a local fishery.

River conditions and bug activity is where the boots-in-the-water approach can be beneficial. With guides on the water every day, they keep notes on the countertop conversations of regular customers. They get it. While often specific to seasons, the reports won't always vary, but it doesn't hurt to get the information when the summer shifts and the underbelly color of the hopper changes from purple to red.

The strong retail outlets will have active online reporting. It isn't difficult to pick out the ones that have timely, well-written updates. These can be collated or even saved into a Word document for future reference. With simple copy-and-paste, you can save years of reports. This can allow for better seasonal breakdowns and tracking of when and where these rivers change. If you want to make the day of the shop employee or owner, reference these notes on the next call. You'll be amazed at how appreciation of someone's work can build rapport. You may just get a little bonus nugget of intel next time around.

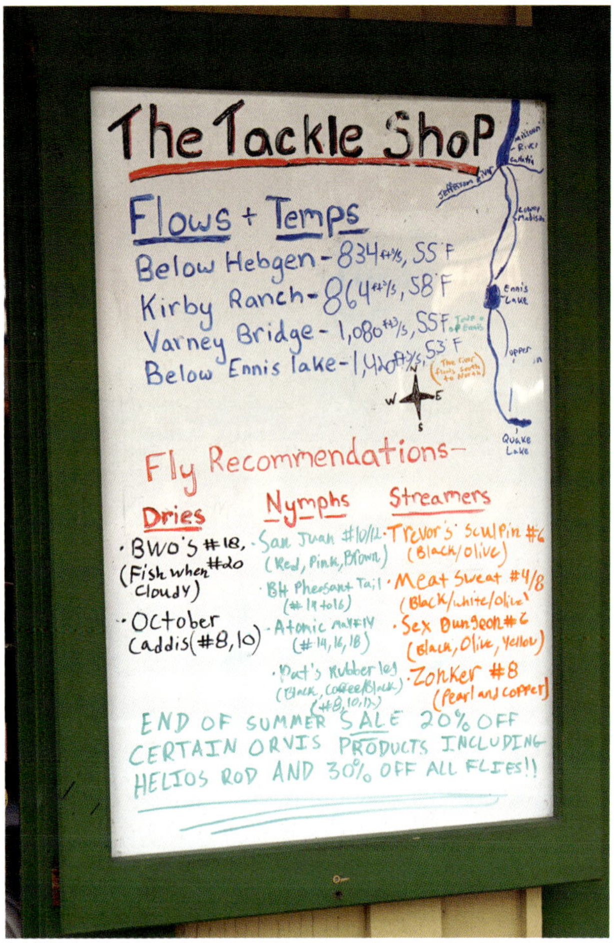

A quick walk through town can provide current river and hatch information from a simple whiteboard. Local shop reports provide a more real-time view of bug hatches and cfs. Simply walking by the shops can often provide more localized intel. PHOTO BY CAPT. COLLINS D. ILLICH

After a few weeks of reports, the quality data should float above dated, vague informational posts. Make notes about which shops post most often and how they present the information. How often do they reference water flows (valuable information available from the USGS, which can be a point of check and balance), and what times of the year do they post? Is it a seasonal copy-and-paste of the same fly patterns? Or do they offer some color to their fly suggestions?

The second avenue to consider is a hunting app. While fishing-specific apps can be great, to go a step further, we suggest investigating more hunting-specific apps. These can be valuable sources to better understand property lines, elevation changes, and access points. The expanse of topographical maps and routes not found on common fishing apps can greatly improve how you view a given river. It will often lead to creeks and high-mountain lakes. Remember, hunters have to hike a long way while carrying gear and trying not to spook a wary animal. We as anglers may have it a bit easier since our target is hiding in the water, but the ability to study a map and take a clear and safe route is similar.

As hunters face an increasingly troubling challenge to access, consider their struggles and need to understand antiquated rules and regulations the same as yours. We all face shrinking opportunities and tightening of the belt from landowners. Not only can the study of this area keep you legal when stepping on a stretch of water, but it can also teach you a unique way to view access opportunities.

TroutRoutes, a fishing-specific access site app, has come to the forefront while presenting not only access but also local resources for anglers. It intends to allow anglers to explore confidently with insights and knowledge of where they stand. According to the website, TroutRoutes also touts the ability to connect anglers to local resources and experts, such as fly shops, guides, local brands, and nonprofits. So, unlike other apps that focus heavily on the DIY aspect, without providing deeper information, this app will provide resources on over 50,000 cataloged access points. As is true with all app sources, there are fee structures and levels of participation from basic maps and waterways to more detailed and advanced-level access. But for new or infrequent visitors (and those just looking to explore), the investment can be worth it.

All of these tools can greatly aid the wading angler in finding access, and by combining digital resources with functional fishing and casting tools, we believe a vastly more successful angler will emerge.

How Most Anglers Approach Wading Water

Floating anglers have an advantage or are unwittingly tricked into fishing new water as they meander down a river. Wading anglers are not afforded that freedom of rapid movement. They don't have the benefit of "quickly moving on" if they disturb a run or make poor fly placement. So, as a wading angler, an elevated fishing IQ is demanded. While this may seemingly add challenge, waders should utilize patience and observation to maximize an opportunity.

Eight out of ten anglers will head to the most welcoming water in any run first. It's human nature. They often gravitate toward the water that they have previously had success in. Typically, this water is dissected too quickly, with the angler placing a fly in the most obvious areas. The choice water is fished first, when likely the closest seam and slack water is ignored. The same typically goes for the water in front of and farther behind an obstruction.

There are some bright red warning flags to watch for when observing another angler. We will use a common water feature, a pool behind an obstruction, as an example.

1. Anglers quickly fish the pool or deeper, slower water immediately behind the obstruction.
2. They move through this water too quickly (or, conversely, not quickly enough).
3. They don't focus on the water above the obstruction.
4. They forget to fish inside and then out.
5. They start in the middle of the pool and don't fully work the length of the water.

Devin Olsen further discusses this approach and how the mental flexibility of most anglers is nearly nonexistent when it comes to familiar water. In his video with angler and tier Lance Egan, *Modern Nymphing*, Olsen posits that if fish don't strike on the first or second drift through the most obvious water, it is time to move on.

As detrimental as fishing a run too quickly, the cement-booted angler can burn too much time and energy on unproductive water. Equally common, and something that has been ingrained into the psyche of all fly anglers, is the notion that if you just fish it more, make a few more passes, a hookup will occur.

Both scenarios will typically place anglers in the middle of the pool or directly beyond the obstruction, with limited movement. The intrepid angler, maybe with only three to four casts, should move both above and well beyond the boulder. The ability to target multiple areas more effectively will lead to more success.

Finally, the angler who approaches the above water keying on what is close before stripping the line or moving out puts themself in a much better position to succeed.

Fish What's Ignored

Most anglers hit or cherry-pick the most productive-looking water and quickly move on. How and when they determine what is "productive" can be quite finicky and oftentimes is based on what worked before. This assessment is frequently made without the full history of seasonality, temperature, and time of day. Many factors can influence productive feeding times for western trout.

The best way to articulate this would be to break down one of the most common water features in a river. A large boulder and pool water regime is likely the most common feature encountered in the West. In this scenario, most anglers will enter the run halfway between the boulder or whatever is creating the disturbance/change in water flow. They will flop down a few casts, right behind the boulder. In smaller, low-flow water, this is typically the least productive position. Water will tend to eddy or cycle back toward the boulder, creating awkward fly movements. Remember the path of least resistance. While it may reek of laziness, the less energy expended by a fish and the more calories it can inhale, the better.

After working close-in water and starting well above the boulders, the angler here works her way downriver. While it would have been very tempting to just step into the middle of the pool and make an obvious cast, she instead decided to work the entire stretch in a methodical manner, which yielded many landed fish. PHOTO BY NICHOLAS CONKLIN

Small, easy meals are often more appealing than an exhausting chase at a big meal.

Skilled anglers focus on either side of the obstruction and the slower water in front of and well behind the obstruction. Depending on the structure and water scenario, only a rod length distance from the bank may be the perfect holding line for some trout. Always cast to the water directly in front, before extending the line out. There is a surprising (but really shouldn't be at this point) number of fish that prefer the slower water that is often found just feet or even inches from your boots. Anglers often miss this combination of mental and physical approach to this type of water.

Whether it is verbalized or not, they think, "No fish is this close!" They assume that based on the time, money spent, and general distance traveled, it would be almost cheating if they hooked a fish three feet from the bank. There is no realistic correlation between the two ideas. If you don't believe it, next time watch the local kid catch fish on his cheap push-button outfit while dragging a Mepps spinner a foot from the bank.

This is a common fallacy with anglers casting and fishing longer, higher-weight rods. Long rods can be great line-repositioning and fly-management tools. They are not solely about bombing long casts and covering the faraway water.

Ego and a perceived sense of power will drive anglers to strip off too much line the instant they step into the water. They feel they need to touch the other side of the bank, often disrupting and blatantly ignoring the opportunities just feet away. Just because the technique requires a longer tool doesn't mean the basic approach of fishing close and then extending becomes null. Streamer and swinging anglers are most aggressive when it comes to fishing too far out.

As mentioned above, the way an angler moves and how those body movements translate into disturbance of the natural environment will play a role. By plodding into the current or not taking appropriate time to observe a section of water from a distance, anglers will inadvertently move fish out of these close-in, comfortable holding lies. They spook these fish, pushing them out of a comfortable lie, and then end up standing in that spot fishing. This results in dragging flies over already alerted fish or into a run that may have been spooked by the fish you just bumped out of its lie.

While not the most easily accomplished when fishing a heavier or sinking type of fly, this is an area where dry-fly anglers can pick off several fish without disrupting the larger pool. Willing targets can be found just inches from the bank, and with an appropriate upstream cast, a greater percentage of the pool can be fished when starting close to the bank.

Another overlooked portion of this water regime is the area of flow above the obstruction. Whether it is difficult to wade or it is assumed the water is moving too fast to hold fish, anglers quickly glance past this area. Anglers start their drift several yards above the boulder, searching for fish in holding lies or resting spots after working through the fast-flowing water. This can be a great spot to hook a less pressured fish.

Once a fish maneuvers ahead of this obstacle, it will need a spot to relax or pick up calories before moving on. Additionally, when fishing an area strewn with boulders or other obstructions, this is a common "drop back" position for fish to hold if they had been pushed downriver by that clumsy angler upriver from your position.

Wading Solo or with Others

When wading with others, you usually have the advantage of carrying fewer rods and less individual gear. More important is the ability to quickly define what is working and what is not. And an effective team will adjust quickly.

Where the team angle comes in is when you know the hatch and type of bug that is going to work, but you want to maximize efforts and refine the size or tint of the fly. Working through a river section, you

Fish will often prefer the slower water that is frequently found just feet or even inches from your boots. Here the angler started while up on the rocky bank, before carefully stepping into slack water. With this being the first spot of the morning, she made several thoughtful, short casts until she got comfortable and felt she had presented a fair offering to the close-in water. PHOTO BY NICHOLAS CONKLIN

On a chilly Montana morning, two anglers begin by working smaller sections of water together. They each took one rod, with two different styles, working until they discovered the effective approach. PHOTO BY NICHOLAS CONKLIN

may send the first angler through with a larger or flashier fly. Then the second angler works through, with a nymph, dry, or streamer in a smaller size and more drab color. If this process works, repeat it. If not, you have reduced time in inactive water.

If you pick up a few fish through the stretch but suddenly hit a wall, you have a few options to trigger some aggressive strikes. Shifting to a large, gaudy fly and working it confidently through a run or pool at different angles can reengage a shy fish. Consider fishing the streamer down and across hard. If no strike occurs, walk toward the tailout and cast to the head; by quickly moving your fly downstream (back toward you and the tailout), you may elicit a strike, as the fly will be moving with a cadence that most anglers don't impart to a fly.

In addressing the components of wading with other anglers, we will break it down into two parts: the competence of the angling "team" and the approach when on the bank. You could consider both as outlining clear expectations. These initial conversations should be covered, whether with a longtime friend or a professional guide. Anglers, new or experienced, should not shy away from verbally defining goals and expectations. This is also the time to openly cover experience and skill levels. No guide will fault an honest conversation before a trip.

Years ago, during a group week near the upper portion of the Madison River, we were fishing with anglers of varying skills and wading experience. We had been switching back and forth between floating some sections and wading fast-moving, rocky stretches. Unbeknownst to us, one of the anglers had, with best intentions, removed the metal bars on the bottom of his wading boots in order not to scrape up the bottom of the drift boat. However, doing so reduced him to wading with what was essentially a sole as slippery as a cowboy boot. While incredibly unsafe, and likely difficult to maneuver on the slick, irregular rocks, he did manage a few fish and took only one small tumble. While likely rare that more experienced anglers will encounter this issue, the clear communication component is important. When planning to wade with others, regardless of skill, it's important to communicate expectations and understand angling limitations.

What is the plan of the group of anglers before and when on the water, and how can they articulate their individual goals to ensure success? This level of communication should proceed as new water types are approached.

In one typical scenario, two wading anglers have approached a small elbow on a large western river. One angler started high above the riffles, with a plan to work down the river, utilizing a high-stick or Euro-nymph-style rig. The second angler, after noticing a few rises along the deep-cut bank, is working the river with a double dry-fly rig. Typically, an angling team would have one angler work down while the other worked up the river. As an experiment, however, they decided to start in the same section, with vastly different techniques. This was done to cover typically high-pressure waters.

After watching a few anglers either pass by the section or quickly work a hopper-dropper from the above bank, they decided to take a quick but diverse approach to this section before moving along in opposite directions. They proceeded to fish the next 100 yards of the river, in different directions in a comparable manner, with success.

The evening before, they had sat down at a nearby bar, with a river map and a hunting app. They picked a starting point and marked off the distance from the parking lot (it was only around a half-mile from the ramp, which was intentional). They discussed their goals, writing them down on a bar napkin:

1. We need to fish the overlooked, poorly targeted water near the parking lot.
2. We need to start here and then head in opposite directions, working the smaller pools and cut banks.
3. One angler will fish a surface fly.
4. The other will fish subsurface.
5. We will fish in opposite directions to this point (or distance) and then meet back in the middle.
6. We will reevaluate fly selection and adjust based on fish response.

While a somewhat clinical approach to fishing, this method paid off for the anglers and resulted in quality fish. What it also accomplished was an actionable plan to approach unfamiliar water.

Casting before the Wade

Chapter 5 discussed casting from a boat or raft and how anglers should apply many of those practice techniques to wading scenarios. Of course, angles will change. Relation to and distance from the water and to targets will take some time to become acquainted. However, by utilizing your lawn or another grassy area, you can get some effective practice. Use household objects, cones, shoes, sports balls, or anything you can set up on your field to simulate a stretch of river. Again, you can tie up a mock leader system and faux-fly weight to become comfortable with your distances.

We want to reiterate the need to practice casting off both shoulders and casting with both hands. Now is the time to play with your cast. It is also the time to slap yourself with the line, pile it at your feet, or snag the trees. Don't wait until you're on the water to discover you haven't practiced enough.

As a wading angler, it is important to frequently reacquaint yourself with the head length of the line and the distance you will be fishing, as this will change frequently. This can be easily practiced on dry land with the placement of markers. Stripping off only enough line to hit your target, hit the mark on a few casts and then pull off another few yards. Moving farther out, continue to practice line management. Starting close, slowly work out lengths of line depending on your target, and, as you are doing that, you will be able to stretch and pile your line in the best manner to mitigate any tangles or knots when casting.

Once targets are hit consistently, consider working through the same course while on your knees. Keep in mind that your wading depth or position on rocks and boulders will change, so don't become static in your practice. This is also the time to get reacquainted with the fly line and rod. This is one key area where the wading angler will have the advantage over the floating angler. The wader will not be hamstrung by the skills of the rower or the limitations of the behavior of the river.

Not only will these practice steps increase casting and line management proficiency, but they will also help calibrate distances and targets for when you are on the water. By pushing yourself relative to casting angles, positions, and line management techniques

A bright orange practice fly line can be a useful tool when looking to sharpen your casting before stepping into the water. Here the caster is doing a bit of practice in the parking lot, hoping to work out the shockwaves in his forward cast. PHOTO BY OLIVER SUTRO

now, you can be better prepared when on the water. An active fishing scenario is not the time or environment to fuss with tangles and try to refine distance to a target.

THE BEST OF WHAT'S AROUND

If you don't have a rolling lawn with ample space to practice casting, consider a local park or crowded public space. By utilizing and working around the obstructions found there, you will feel more comfortable on a stream with brush and trees.

Outside of learning various casts around these obstructions, this will also help work the mental angle of casting under pressure. Background noises such as trucks rumbling down the street or other random noises can act to increase mental focus. Practicing casts with animate and inanimate objects around will help center focus and mitigate pressure when stepping into new water. Believe me, your local park has seen far stranger people and hobbies practiced. Don't be shy, and maybe you will make some new friends.

CLEANING LINES AFTER LAWN PRACTICE

Remember to always run your fly line through a cleaning process after lawn casting. A simple sink bath of warm water and dish soap can help remove any dirt and debris particles and allow for smooth shooting through the guides. There are many brand-name line conditioners and cleaners on the market. It's always a good idea to keep a bottle and clean rag around to apply to the line before and after a session. While a tedious procedure, frequent cleaning will extend the life of a fly line.

Be mindful of clearing away trapped debris from any loop-to-loop connections. This seems to be a common catch point for grass and other particles. The same can be said for the snake and stripping guides. It'll be beneficial to quickly wipe them down before storing them in the rod sock and tube. While guides are now made from some of the hardest, most durable materials, it will benefit you to take that extra step to keep your tackle up to the task. Remember, the goal is to become more successful not only by efficient casting and fishing but also by properly caring for the tackle we use.

One note of caution: Consider trashing your leaders after a few sessions of lawn casting. Don't take them to the river and expect them to perform flawlessly. While it's important to practice with what you will fish with, it's not the best idea to take your fancy new fluorocarbon leader, either tapered or hand-tied, to the park. Instead, do some bargain bin shopping, either at the local shop or in your tackle collection, for a similar/appropriately tapered leader for the technique you wish to practice. It could be older branding and packaging or just a leader that's sat around the bin a few years. While potential sun and temperature exposure may have affected it to the point of not trusting it in a fishing scenario, it'll be a cost-effective material for practice.

Situational Line Management

The ability to manage line or mend depending on approach and technique is a vital skill to master, and it will become more apparent and evolve with time. However, it is often used incorrectly, to the point of abuse. Just because it's there doesn't mean it needs to be used. Or, if you're going to use it, use it right. When we say used incorrectly, we are referring to the execution, not necessarily the need to act.

Also keep in mind that your relative distance from the water when wading, the type of rod and line system you are fishing with, and the distance of the cast will all have an impact on the mend. The same mend and energy exerted when wading will not necessarily translate the same as when you are floating. The style, approach, and rod movements will vary.

Simply put, a mend is a line-repositioning maneuver. It can be done to keep the fly line from outpacing a fly or an indicator or used to help get a fly to sink deeper and have a longer period in the strike zone.

Unfortunately, however, it is most often used as a medieval-era weapon to catapult flies out of the strike zone and cause a nice, garbled pile of line and leader. While any casting instructor will rail against introducing any type of slack into the system, tiny

A higher rod-tip angle and line management kept a clean drift and allowed the angler to quickly pick up and reposition his fly. Much can be learned from the climbing industry when it comes to tricks for effective fly line management.

PHOTO BY NICHOLAS CONKLIN

amounts of slack in this scenario can be helpful. When done to eliminate drag on a chain of flies, a gentle mend and some slack into the line resting on the water can be effective.

While wading anglers will have a bit of a disadvantage here, as movements will be restricted, they can still perform some effective mending. One way to remedy this is an upstream mend. Beyond the upstream mend, focus on where and how you hold the rod tip. In varied current flows, by keeping the rod tip higher and managing the line off the water, you can avoid drag being imparted into the drift. While many chapters have been dedicated to a wide variety of mends and line maneuvering for effective drifts, we pose that as a wading angler, whether you are forced to fish upstream or downstream and however river traffic dictates your approach, there are simpler ways to classify necessary mends.

Not all mends are created equal, and it is important to mentally categorize different mend styles for different approaches. If fishing downstream or down and across the river, your most likely approach will be with an indicator and multi-fly rig, a high-stick or tight-line setup, or a swung fly/streamer approach. Since tight-line anglers won't necessarily require mending, this section will focus more on traditional line and fly systems.

FISH-FIGHTING ANGLES

Whether fishing from a boat or wading, every angler should be aware of the angle of the rod when hooked into a fish. Rod tips are not designed to accommodate any more weight than that of a fly line. When angles are exaggerated, creating an aggressive bend in the top portion of the rod, tip sections are likely to fail.

As a wader, this process can be even more complicated due to the inability to manipulate angles. Ideally, anglers should strive to keep a rod at or below a 45-degree angle in relation to the water. This position allows the most pressure to be absorbed by the lower portion or butt section of the rod. This will allow for effective pressure to the fish while not stressing the tip of the rod.

A consistent, upstream rod angle kept the fish tightly hooked in the corner of its mouth. Anglers should always be thoughtful about keeping the rod tip low, with pressure on the fish coming from the butt (handle) of the fly rod.
PHOTO BY OLIVER SUTRO

The angler begins with a subtle lift to reposition and then will manage the line on the water during the drift. Due to the length of the line needed for the drift, and being conscious not to tangle the rig, she slowly lifts before making a wide arching cast to land her flies back upstream. PHOTO BY NICHOLAS CONKLIN

As a wader, the focus on hooking a fish should be on keeping the fight to either side of the body, maintaining maximum pressure. This method will also eliminate the added pressure caused by current. An angler with a low-to-the-side angle can ensure a solid hook set and maintain more control over the fish during the fight.

INDICATOR RIG MENDS

An indicator rig mend is considered the most difficult to master, and the one most likely done incorrectly by anglers. Subtlety or less movement of the rod tip and, frankly, practice will be key here.

An effective method for naturally moving or jerking flies from the strike zone is subtly piling line on the water. By piling coils of line on the water, you can reposition upstream while keeping your flies sinking at a steady rate and not disturbing the surface object (either an indicator or a dry fly). By practicing slight movements of the rod tip and becoming comfortable with the amount of energy needed to move the line and indicator, anglers will keep the flies sinking and in the zone.

Too much power, and the whole drift can be blown. In the "Casting before the Wade" section, we cover a few tips to become confident in the movements needed to gently lift a line and an indicator during the drift. The boat-bound caster will likely reposition upstream due to float and river speed. Because of their fixed position, the wading angler will need to adapt the mending over a wider array of current speeds and potential points of drag. If casting across a slow-moving section of water, likely standing closer to the bank, into the main channel with more turbulent water, the line is going to encounter drag, causing your flies to move in an unnatural state.

Be prepared to adapt the position or height of the rod tip as well. If drifting across the water at dramatically different current speeds, a higher rod angle can keep the line and indicator farther off the water.

SWUNG FLY/STREAMER MENDS

For the swing or streamer angler, line management is equally important, as the thicker diameter of the fly line is a willing target to grab different current speeds, adding unwanted speed to the drift. What is affected by too fast of a swing is the profile presented to a fish. A slower, steady sink rate, swing, and strip can allow a much more appetizing angle to the fish.

Longtime guide, casting teacher, and overall fly-fishing guru Ed Ward has developed several swung-fly-type mends. His ideas came from fishing heavier flies and sink tips. The key to the "pull back, drop down" (P.B.D.D.) mend focuses on the efficient sink rate of the fly while offering the best profile to the fish. The repositioning of the line allows the fly to sink at a steady rate while preventing the fly line or shooting head from creating drag. Slack can create too much sink, followed by a sudden rush of movement as the fly catches up to the rapidly swinging line. Both do not allow for the proper time and profile for a fish to move on the presentation.

During the P.B.D.D. mend, an angler swings the line out while subtly dropping the rod tip to reduce the amount of slack in the cast. As the current continues to pull the line, the fly continues to sink at a steady rate. This eliminates the need to make an often exaggerated, aggressive mend that rips the line and flies out of the zone, interrupting their sink rate. The flies get jerked out of the strike zone and the profile changes (consider this point when contemplating mends on indicators).

DRY FLY MENDS

Delicacy is key with any surface- or dry-fly-specific approach. True delicacy can be achieved only through precision. Precision is achieved through experience.

On a recent trip to a southern Colorado river, during an immense Trico hatch, the art of subtle mending and line management took even a seasoned angler some time to re-master. After a dozen or more casts and quick drifts, it quickly became apparent that I (Nicholas) was (1) fishing too much line and (2) not managing the drift as effectively as I should have been. So, to more precisely allow the fly to drift into the sipping lanes, I made two corrections.

An angler working different currents uses both her line hand and her rod hand in order to mitigate as much drag as possible. Maintaining an active line hand will allow her to prevent any slack caused by the water currents and keep the fish from coming off the hook. PHOTO BY OLIVER SUTRO

Angler Jason Randall presents a subtle, aerialized line mend to get a realistic Trico drift. These mends will take practice, but with subtle rod tip movements, while the fly line is in the air and unrolling, anglers can better combat varied currents. PHOTO BY NICHOLAS CONKLIN

First, I shortened the amount of fly line I had outside the tip of the rod, lengthened my leader, and started to perform more subtle changes to the path of the rod tip. The line and leader change allowed me to focus on keeping the larger, more drift-resistant part of the fly line off the water, and I was able to manipulate the thinner leader more effectively. I also discovered that while using a delicate technique, I could reposition much closer to the feeding lanes. The fish did not notice my presence as millions of small bugs drifted by.

The second step was to perform a reach (or aerial) mend, allowing the fly line to land more upstream of my position. This prevented the line from bellying down and dragging the leader and fly. In this scenario, I was casting at more of a 45-degree angle downstream, allowing the flies to drift freely. With a subtle reach and movement of my rod tip during the forward cast, I was able to drop the fly line in a rounded shape, keeping it from getting pulled quickly by the current. Similar steps can be taken when approaching from the more traditional dry-fly route.

Anglers casting upstream can benefit from positioning the obtrusive fly line to the side, allowing only the tippet to drift past a feeding trout.

Body position relative to the water can also be a quick way to better present a surface fly. As experimented with during the Colorado Trico hatch, moving farther out had no negative impact on fish behavior. While books and educational materials throughout history have likely instilled some specific, unbreakable rules to dry-fly fishing, there are none. The fish don't care. I don't think they bother to read most fly-fishing books. The Gen Z trout are more focused on what's right in front of them.

If specific mends and general line management are difficult to achieve, or you find you're creating too much drag while fishing across varied water speeds, consider adjusting your physical angle. By making small shifts in footing and general position

Dry flies are challenging but very rewarding. With effective mending, this Colorado trout felt the fly and its movement were realistic enough to gamble on. PHOTO BY NICHOLAS CONKLIN

in the river, anglers can get a more effective drift and better presentation.

Don't be afraid to move, shift angles, and, especially with a dry fly, get close to the target to make shorter line presentations. If the fish are not wary of movements, it is wise to shorten the distance the tool needs to perform. Ultimately, the end experience will be either enhanced or diminished by the quality of the drift.

Intersection Zones

With the rise in large, multi-articulated streamers and popularity of tying, buying, and fishing large flashy streamers, most anglers have not quite grasped the point. While it may be cool to drop $25 on a flashy, furry streamer to stick on your dash as you drive around or add a "meat chucker" to your tailgate or Instagram bio, the coolest anglers have gone a step further to understand how these large, water-displacing flies can be productive when fished in intersection zones.

This is where wading anglers and those drifting downriver in a boat need to cross-reference notes.

For the wading angler, these intersection points do not always have to come at a 45-degree angle downstream while beating the banks. These casters can hit different angles and focus on where a large predator might attack a fly. Fish may tend to get turned off by the constant profile afforded here as anglers repetitively present flies.

While we referenced a large streamer to highlight the convergence of these zones, do not mistake that style of fly as the only option. More specifically, we suggested this fly as a "searching" pattern so that anglers can more clearly see where the differing currents join. By watching the movements of a large fly, anglers should then be able to better approach this water with smaller flies and delicate presentations.

Perhaps the fish has been pushed out of a safe pool and has gotten swept up in the faster, swifter current. It is desperate to get out of the main flow where larger fish are transitioning and seeking quick safety in slack water. As it works its way toward the corner of the "triangle," it is intersecting with a larger predator holding just on the cusp.

When looking for this triangle, or funnel point, do not always assume it has to be large or otherwise obvious quantities of water merging. Even a small trickle from runoff may be enough water to help direct bug life.

Still Waters Run Deep

Several years ago, while working at a fishing camp in Kamchatka, the idea of safe and slow wading never became more apparent. The first group of the season had just arrived and began fishing within eyesight of the camp. While waiting for other guests to arrive and a welcome happy hour, a client eagerly put on his waders and charged into a large freestone river. After hooking a handful of fish, he decided to make his way back across the river.

While sitting on a bank just down from an Alaskan river camp, acting as a guide, I (Collins) had a clear view of the varied river bottom. What the returning, excited angler did not notice was that he was on a straight path toward treacherous water. In the excitement of arriving on his trip and lucking into a few fish, he had meandered just a few yards downstream from where he originally stepped in. However, the water could not have changed more drastically. Despite vigorous yelling and arm movements to warn my client, he ended up past his waders.

While likely funny once he returned to shore, the serious nature of wading too aggressively was apparent and can quickly put an angler in peril. By ignoring a safe speed and approach, this angler could have forced a premature end to his trip and even life.

Whether approaching new or familiar water, anglers need to always consider their steps. How do other anglers walk in? At what angle are they approaching their "target"? How are they approaching the boulder—do they fish close and work out? Or do they barrel head over heels, flop down a lame cast, and stomp down the river? Most observations

Careful, thought-out wading steps will combine safety with a more precise approach. This should be practiced and performed both in and out of the water. PHOTO BY OLIVER SUTRO

Although some rivers don't appear to be deep, the speed and mass of water moving these "midsize" rivers can be deceptive. Here the angler takes an upstream path and focuses on keeping the lead foot forward.
PHOTO BY NICHOLAS CONKLIN

will likely lead you to one conclusion: Careful, thought-out steps can lead to success. They can also be paramount to safely navigating a stretch of uncertain water.

Wading anglers are at a distinct disadvantage in most scenarios. They simply cannot cover as much water as a floating angler. Conversely, you can use that as an advantage to shrink your vision and more specifically focus on the stretch of water in front of you. Take your time, and thoroughly dissect a given portion of water.

When we suggest "shrinking" a section of water, this can be as large as 20 to 30 yards to as small as 4 by 4 feet. By massively shrinking down a section, whether it is behind a boulder, a small pool, or a section of riffles, this microscope approach can force you to understand a river. These micro sections, whether you run a fly through them or not, are a great examination mindset to adopt.

Spend deliberate and thoughtful time on one small section. While a floating angler can have several river miles and holding lies looming over them throughout the day, the intimacy a wading angler can generate with a stretch of river can become a strength. Consider making a mental note of some physical objects along the water. These can help funnel your sightline to focus on a water regime before moving along. Fish it thoroughly before moving toward your next spot.

By watching how water flows around obstructions, and watching the water move, you can learn a lot. Are fish holding here? Or does this seem like more of a transition zone that they may pass through? After some observations, make a few passes with your flies before moving on to the next section.

If your section has some varied flows, instead of disturbing them with flies, consider dropping a natural object into the water to better understand the current speeds. Although the water may seem like it is flowing consistently either vertically through the water column or horizontally, it likely has some change. Acorns, small sticks, or leaves are natural objects that fall into the water every day and will have less of a startling impact on the fish. This type of grid fishing will not only maximize your time but also help create a more thorough mindset when approaching heavily fished water.

Consider the approach of Jason Randall in his book *Nymph Masters*. Randall has some highly scientific yet incredibly commonsense approaches to understanding how varied current flows. He breaks down various river sections and uses underwater illustrations to highlight just how much of a speed difference there is and how your flies can travel at drastically different speeds given their size and weight. The ability to slow the pace and analyze a path before stepping into the water can be just as important as having the right fly at the right depth.

In addition to considering your path to fishable water, keep in mind the return trip. Many western rivers have very deceptive flows. These turbulent, powerful rivers are difficult to wade. Powering out into the water will burn energy and can lead to trouble when it is time to wade back to the bank. Changing light conditions, flow levels, and energy can put wading anglers in unsafe conditions. Develop a plan and a set timeline in uncomfortable water where you are not in tune with the potential for change. Safe wading is as important as the right fly or perfect stretch of water.

Knot Tying and Rigging

Knot tying or, better said, the ability to perfect knots while standing knee-deep in a swift current is another item for the angler looking to improve. Not only are a handful of precise knots crucial to fish-landing percentages, but knot-tying also accounts for wasted time without a fly in the water. You've spotted it while driving along a river or standing downstream from a fellow angler—the hands together, fumbling, rerigging look.

It takes most anglers a significant amount of time to rerig and retie while in the water. But there are two methods to consider to improve efficiency.

The first method is to maintain your position in the water and have one or two knots you can tie with your eyes closed. The second is to ease out of the run, wade carefully to the bank, and assume a position where you can "block out" the environment and quickly rerig.

What's important here is the tying practice. One method we suggest is during your non-fishing hours, practice knots in some unconventional ways. For example, with the "walk and tie" method, you walk around your home or outside while tying a length of line onto a multi-fly rig or attaching two chunks of leader material. Moving around can help better focus your attention on the work at hand. This can eventually help alleviate the struggles and anxiety that can arise from having to retie in midstream.

When you feel you can comfortably maneuver around your lawn, while working a minimum of two knots in a time-sensitive manner, the next step is to try it in your wading apparel with a fly rod tucked under your arm. Never mind the neighbors peeking through a slit in their blinds or what the local kids may say about the weird person who paces the yard, dressed in fishing gear and practicing knots. This type of practice again alleviates the anxiety of retying midstream, while also providing a semi-realistic dress rehearsal of sorts.

One of the most challenging parts of this equation is the handling of the rod and reel. By doing this outside of the water, you can switch holding positions and gain a different level of comfort, so when the time comes to maneuver the rod, reel, line, and flies while standing in the water, muscle memory will return, and the entire process will be much more efficient.

Tackle for the wading angler can be a tricky road to navigate. Carry one rod or two diversely rigged pieces of tackle? These decisions may help drive Reddit forums and internet searches, but finding the right tools for your most fished type of water will take some introspection. PHOTO BY NICHOLAS CONKLIN

CHAPTER 7

TACKLE FOR THE WADING ANGLER

We could go on at length about the constant evolution of tackle and the great advancements in composite materials and components. All of these have an impact on performance—this isn't an advertisement for the latest and greatest tackle. Despite just consuming more products and hoping they will increase success, we want to propose developing the skills to fish the tackle you already have, but better.

Let's start with the rod, which is the most important part of the equation, and one that can present the highest margin for error, or the all-too-common errors that arise when anglers select a rod. Whether it is from a skewed sense of capabilities and strengths, ignorance of casting faults, or the advice of others, many initial casting struggles can arise from selecting the improper rod.

Is it the right rod for your casting style and experience level? Despite what your local shop staff will tell you, fly rods are as unique and specialized as the individual. Every fly caster has a different tempo, casting stroke, and style. As a sport and skill that can be either improved on or neglected, it is important to remember that different tools will be required as the caster develops. *Development* is a key word here, as that progression is not always linear. Interests, time spent participating, and several other factors over the lifetime of the fly angler will impact casting style and ability. These things evolve, as should tackle choices, starting with the most important part of the equation: the rod.

One of the reasons manufacturers make so many different rods is not simply to sell more and increase the bottom line. Whether acknowledged internally or not, a variety of rod actions and levels of power (stiffness) are key in offering a well-rounded lineup. Anglers get beaten over the head with the term *fast*, but it's often not the best action given the angler's skill level or the targeted species and technique. While many anglers develop their casting stroke and tempo and graduate to a tool that might be faster and stiffer, a significant percentage of anglers stick to a more moderate-action rod. While no statistical data exists, odds are as they age and progress, many tend to prefer slower and softer rods.

The end goal should be to provide the right tool that fits the angler. So many rods in various actions and levels of stiffness or softness exist. It lands on the angler to judge for themself what feels the best. It's still a world of personal preference, where no modern, quick-cut video or fancy social media ad can adequately place the correct tool in hand.

While some companies have started to market to their audiences more effectively and better describe rods that are skill improvement tools, most still hang their hat on the latest spiral weaves of nanofibers, high-modulus X-materials, space-age materials, and overtly techie talk. Without a standardized scale for describing the level of stiffness or softness in a rod, that component often gets confused with the action of the rod.

What matters here is feel. Unfortunately, there is no modern "how can I buy my way into a good cast?" solution to this problem. It is time spent test-casting and playing with rods. It is not a skill you can earn by walking up to a rod rack and forking over however much money is perceived to make a caster great. It takes time, correct practice, and a desire to find the right casting tool.

Rods do not inherently aid in providing an angler more accuracy or distance without skill. Alas! The local shops, events, and consumer shows have

The rod action will depend on several factors, but the "feel" will play more of a role in fitting the caster's stroke. This really can be done only by casting a few rods, with various lines, at local fly shops and consumer fly-fishing shows.
PHOTO BY OLIVER SUTRO

Casting, just like fishing, should be an exercise in experiments and trial and error. The angler here has spent considerable time developing the ability to precisely land a multi-fly rig and indicator. PHOTO BY NICHOLAS CONKLIN

reemerged in significance. While we thought they were just dozing on Main Street or in some dim and dusty convention center during the cold months, they were secretly waiting for their window of opportunity. Getting to a physical venue and casting several rods is truly still the best way to gain knowledge of what works for you. It can also highlight what doesn't feel right, despite what you have been told or felt inclined to believe after hours of online research.

After spending time finding a rod that "feels good," the next step is learning how to use the rod correctly. And that means casting practice. Starting to see a theme here?

Learning some tricks and completing training practice, to help make that old 9-foot 5-weight of yours perform just a little better, will help. Casting, just like fishing, should be an exercise in experiments and trial and error—an ever-evolving learning process. But, unlike the Pythagorean theorem and solving for *y*, this process should be enjoyable and will be useful later in life (or at least in your angling life).

The ability to re-create realistic fishing scenarios and apply them to practice casting sessions will provide a critical link to separate one angler from another. Seeking out professionals is another area where great strides can be made in efficiency. When we say professional help, we don't necessarily suggest meeting in a church basement once a week, drinking stale coffee, and admitting to deep-seated, antisocial proclivities. You already fly fish, so most academics would have tagged you as outside the norm of acceptable human behavior. Yes, mothers pull their children across the street when approached by someone clomping down the sidewalk in waders, with a raccoon tan and muttering about marabou and flash. However, there is still hope. And that hope comes in the form of professional casting instructors.

Whether they are based out of your local shop or through Fly Fishers International, help is available. It's okay to admit it. Even if you think you don't need it, you probably do. What's the harm in one session? Odds are the coffee is better, the people are more pleasant, and the stories about line breaking on monster fish won't be as gut-wrenching as the ones told at the other kind of meeting. Many resources exist, and with great strides in educational organizations such as FFI, Project Healing Waters, Casting for Recovery, and the like, a lot of skills can be learned, negative traits can be removed from your casting, and new fishing improvement ideas can sprout.

When the foundation is set, and you have the proper thought process to analyze and correct your casting, you can begin to work through the entire system and analyze everything, from reevaluating reels and line systems to improving on terminal tackle, flies, tools, packs, waders, boots, and other gear. Focus can then be solely on the real-time situations faced on the water, and then the whole picture will come together. Everything an angler wears, ties, or waves in the air should be analyzed, examined, and then removed from the equation, replaced, or refined in terms of usefulness.

In the end, all these items will be packed up, strapped to the body, and carried all day. Beyond the physical weight consideration, think deeply about whether that extra zinger, backup nipper, or spare thermometer battery really will add value to the fishing day.

Everything an angler wears, ties on, or waves in the air should be analyzed. The questioning angler is the sharp angler. Find the tools necessary for your fishery.
PHOTO BY NICHOLAS CONKLIN

Multiple Rods

As a walking angler, having multiple rigged rods can be challenging. However, the hassle can be the same if you carry one rod and constantly encounter scenarios requiring rerigging. So, the solution is not always carrying fewer rods. That said, here are some thoughts on what and how to carry multiple rigs when wade fishing. Depending on flow, bug activity, weather patterns, and seasons, all of the following can provide the critical link to improved success. Once the action and rod power best suited to the angler are determined, the competent can drill down to which type of rod is best suited for a given technique.

THE HOPPER-DROPPER, DOUBLE-DRY SETUP

This setup can be performed with a standard 9-foot 5-weight. Action and stiffness are the angler's preference and not a topic we will touch on here. We will, however, discuss how the performance attributes of some rod flexes can be more beneficial than others. If financially feasible, this is where an angler should seriously consider adding a heavier-weight rod to their arsenal to accommodate the larger-than-average hoppers and multi-fly rigs. A 6-weight rod can allow for slightly more mass in the line to deliver those larger foamy imitations.

In most scenarios and most seasons in the West, the double-dry, double-nymph, or dry-dropper setup will be the most versatile. Whether the angler utilizes a standard 9-foot 5-weight rod or graduates to a 6-weight, the ability to fish these multi-fly rigs will be powerful. The addition of another 20 to 40 grains in the head portion can help aid the angler in moving the more complicated leader and flies. This system also eliminates the traditional indicator by utilizing a larger, floating fly as the indicator. While you may remove a nymph or other sinking type of fly from the system, you can provide another attack point for a fish, and one that has a hook on it. Of course, the extra grainage of the fly line will be more than adequate for the indicator double-fly rig scenario. As a mobile angler, this setup aids in the ability to fish multiple techniques and flies throughout the session without transporting multiple rigged rods.

Due to the nature of this style of rod, anglers can also utilize many of the below techniques with one rod. Smaller streamers, dries, and even some rudimentary high-stick approaches can be accomplished with the standard 9-foot 5- or 6-weight rod. It can be done. What does need to be acknowledged is that this is not a magic bullet. It's not the Lee Harvey Oswald of tools. While it can seem like the Swiss Army knife of fly rods and allows an angler to fish many techniques with minimal tackle, it does have its disadvantages.

Rod length and power are limits in the equation. While ideal for several techniques, this rod will not be sufficient to deliver large, weighted flies. This style of rod will also limit the angler's reach and ability to perform the more nuanced European or Czech-style drifts.

HIGH-STICK/EURO NYMPHING

The high-stick, or European, style of tight-line nymphing has been all the rage these last few years and for legitimate reasons.

While in no way new, public awareness of the technique has grown drastically due to the highly effective nature of the approach. These longer, lighter, and highly precise rods have afforded anglers the ability to dab or dump small, lightly weighted flies to precise targets. Spawned from the world of competitive angling, many unique styles of rods, long leaders, and flies have developed in countries

By bumping up in rod weight, more mass in the fly line can deliver large foamy hoppers, streamers, or multi-fly rigs. The key here is to understand the mass, or weight, of a fly line and how it acts to load the rod so that it can be unrolled and deliver the fly. PHOTO BY NICHOLAS CONKLIN

Light lines, sparsely tied flies, and long rods allow for more effective reach, with minimal drag. This rod style eliminates the need for a cumbersome fly line and strike indicator. PHOTO BY OLIVER SUTRO

like Czechia, France, Spain, Belgium, and other European countries. The goal was to deliver an incredibly light-line, stealthy fly system when fishing technical waters. This style eliminated the need for a cumbersome fly line and strike indicator.

Long rods in the 10- to 11-foot range allow better reach, better contact, and more success. For many reasons, anglers have taken to these longer rods and exceptionally fine tippets to target skittish, heavily pressured fish. As opposed to the floating angler who needs to consider how long of a rod to fish out of the boat, the wading angler can benefit from a longer lever.

In most Northern Rockies rivers, we tend to yield to the longer, heavier weights due to fish size and current. Having a bit more power in the butt section can help in controlling a fish and not overly stressing the tippet. Modern 3- and 4-weight rods, weighing less than 2.8 ounces, have increased the effectiveness of this style of fishing. Shorter and lighter rods, such as a 10-foot 2-weight, should be in any Southern Rockies angler's arsenal. This approach can actively combat pressured waters and ideally open new fisheries where more traditional techniques are cumbersome and ineffective.

The ability to quickly position and reposition flies is improved with these longer rods. Without the traditional fly line and need to have a specific length of the head outside of the rod tip to load the rod, the angler's motion is compact and aggressive. There is no need to haul the line or spend extra time false-casting. This is a precise pick up and reposition method that allows the maximum amount of time for flies to be in the water. Most Euro rigs will have some form of level running line or mono rig that does not change the taper, so anglers must approach the casting differently, with the mindset to manipulate a much shorter, albeit precise section of line. This thin material allows anglers to avoid a denser, drag-generating line. Not only does this help refine the ideal drift, but it also plays less of a role when in windy conditions.

Notable is the wading depth, or lack thereof, required with the longer rod. While the floating angler may have the advantage of getting up above obstructions, the wading angler can utilize this stealthier, lower-to-the-water approach to their benefit. The ability to hug the shore and fish close-in water, while not getting your waders wet, only supports the goal of the approach.

Nymphs in various weights and color patterns are nice, but a more thoughtful box with few trusted flies in diverse sizes and weights is more effective. But don't be fooled: Many of these nymphs, whether labeled a Euro nymph or not, can be effective on various line and rod styles. PHOTO BY OLIVER SUTRO

Consider on-the-water obstructions: boulders, downed trees, and other areas for concealment. In most scenarios, your standard 9-foot rod may not give you the reach, fly placement, and wading safety that a longer rod will. This will benefit you in reach and effectiveness when picking up and repositioning flies. It also allows you to be more precise in movements and general stealth. Again, we want to stress safety when wading. A longer rod can prevent anglers from wading beyond their boots and keep a realistic distance from dangerous water.

Beyond stealthy movements and precise fly placement is the amount of tackle needed for this setup. While more spools of tippet are often necessary, the style of fly may require only one thin box. Flies ranging from size 12 to size 22 are most common, and beaded flies, jig flies, Czech nymphs, and a variety of sparsely tied flies can pack away easily and allow for many variations to be carried. Sighter material and add-on sighters such as neon waxes like Skafars can allow for the simple application of a removable sighter—nothing to disturb the performance or quality of the leader, as the wax material is waterproof but also easily removable. It comes packaged in a tube the length and diameter of a standard lip balm, so it can be packed away in a simple pocket.

While this tool is likely the third or fourth in the quiver, understand that, like all rods, it has its maximums and minimums. All too often we hear customers ask, "Well, can I also do a technique X or fish style Y with this rod?" While it can be a great third or fourth rod to have, if sticking to a very stylized high-stick or Euro nymph rod, anglers must keep within the lane of what the rod was intended to do. More multipurpose, multi-technique rods exist and can be better all-around tools if a tightline nymphing scenario is going to be only the occasional effort.

STREAMER STICK

When selecting a big fly rod, we often suggest the two-rod weight difference. This is where a 9-foot 7-weight with a full-wells grip and fighting butt can be a beneficial addition. With the heavily weighted, more aggressive modern fly lines, these rods can effectively deliver a medium to large streamer with ease. Appropriate for single or articulated flies from

Sometimes a floating fly line and changes in the weight distribution of the fly can compensate for not carrying multiple rods or line styles. This is another area where local knowledge and note-taking on water levels can help an angler focus on what type of streamer fishing setup they will need. PHOTO BY NICHOLAS CONKLIN

3 to 6 inches, on either a full-floating or a sink-tip line, this rod weight and an appropriately paired line can more than deliver the fly to the target. For those building off the multipurpose 5-weight, this offers great coverage and flexibility for much of the western waters.

Whether the angler chooses a full-floating line or some multi-density system will come down to the body of water and the seasonal run-off. Many times in the Northern Rockies during seasonal droughts and periods of low water, a standard streamer-style floating line is most appropriate. The wading component of this system can require something slightly different from the line system required by the floating angler. As intersection points change, along with the need to get the fly down quicker, from different angles, an increase or variance of density is often needed.

While the floating angler often only needs to have a heavier tip to "beat the banks" while swiftly traveling downriver, the wading angler can find some hassle in something too dense. Conversely, in situations where the floating angler only needs a floating line to effectively fish, the wading angler may need something with increased density to get their fly in the proper zone. The lack of constant movement can hinder the ability to maneuver the fly.

Depending on volume and speed of flow, anglers should not fear an intermediate or even short sink tip added to the end of a floating line. This will help get the fly down below the surface. It can also slow down the speed at which the fly swims. We will cover this subject in more detail in the next section, regarding the "swing rod" or two-handed approach. However, a denser tip or line can help slow the presentation, allowing for a better fly profile presented to the fish. This approach can be wise on the economic front both for reducing tackle and for the overall cost of gear. Less gear is brilliant while utilizing the current tackle to the highest potential.

If a floating line is the only option or the option that will suit most of your water, and carrying extra tips or a spare spool is not an option, there are other simple tweaks to get the fly into the strike zone.

Consider heavier flies or a multi-streamer rig. A larger bead-head or cone-head fly, with the same amount of material, can cut through the current and get into the ideal intersection. By adding a second streamer, either smaller or slightly larger, more weight is added, but you can also create the action of a fleeing baitfish. As the saying goes, "A small fish eats bait, then gets eaten by a bigger fish, which gets eaten by an even bigger fish."

Creating extra drama, with added movement and weight, gets flies to the proper zone while adding a sense of frenzy. Like a Bravolebrity desperate for extra screen time, this added, dynamic commotion can further entice an anger-based strike from a trout. While the thought may be to reach for a larger pattern, the added material will slow the sink rate. Instead, consider a more sparsely tied fly, but one with more pronounced weight. Another option to reach depth quickly, but with minimal weight or change to the line, leader, and fly system, is a simple piece of split shot. Attached 3 to 10 inches ahead of the hook eye, this can help drop a more water-resistant, fluffy fly down and add some varied action during the retrieve.

The last step to success in this technique is the addition of a second fly. This can be done by simply trailing a small nymph off the back of the streamer. Many large fish have been caught in recent years on this rig. While it may not present the nymph in the most realistic, natural light, it does impart the feeling of a small bug being rousted from the rocks and floating downstream. Imagine a smaller fish bellying up against a shallow shoal of rocks, knocking a bug loose. This allows for two different target points for the fish. If a trout is uninterested in the chase required for the larger prey, it can key on the disoriented nymph that was just dislodged from the riverbed.

TROUT/MICRO SWING ROD

Those with a predilection for swinging flies, or the disgraced, beleaguered steelheader, can find a useful tool in the trout or micro Spey-style rod.

Previously we covered some of the advantages and disadvantages that this longer rod can have for the floating angler. If long, gentle runs or fast, deep pools are something you encounter, adding an 11-foot, two-handed rod to your lineup can provide some fun and a unique way to present a fly. This technique has grown in the past few years, as some traditional single-hand streamer-swinging folks have looked for another way to present a fly. This method is often very productive for the wading angler.

With this waterborne style of casting, which can be accomplished with both single- and two-handed rods, the fly will spend much more time in the water. PHOTO BY CAPT. COLLINS D. ILLICH

The addition of an 11-foot, two-handed rod can provide casting benefits and added efficiency of a fly swinging various portions of the river. Not only does it aid in line repositioning, but it also allows for the wading angler to hit new water and offer the fly at a different angle and profile. PHOTO BY NICHOLAS CONKLIN

As the parade of boats float by, hitting the banks every few seconds, their presentation is all but the same. Regardless of the ability of the caster, the overall presentation doesn't differ. This is where the swing-style approach can have an advantage. With a cast, swing, and step or two along the bank, fishing out and down, the fish are now seeing prey from a different angle. The swing angler has also opened a much greater field of opportunity while not over-exerting themself. Backcasts and line hauling have been eliminated, and the fly will spend much more time in the water.

This method can be efficiently fished at a distance and with varied floating, intermediate, or sinking tips and flies. Flies as simple as a double Woolly Bugger rig can be highly effective on wider western rivers. Light soft hackles swung on a floating tip and long leader can provide a unique approach during periods of increased bug activity.

Long riffle or boulder fields, pools, and sweeping runs can be covered piece by piece and very effectively fished through. This also allows for quicker movement through a braided set of channels or open water. While swinging a fly through deep pockets will be most common, anglers can more effectively manage lines while moving through different currents. Many of the western two-handed rod fishing opportunities afford an angler the chance to fish both banks and flat runs and any other depression or potential holding lie.

With this style of rod, line management can be simple. The ability to pick up a fixed amount of line and reposition it precisely and quickly is important. Running line management can be performed simply by holding loose coils of line in the fingers. This prevents river currents from dragging running lines or catching, as occurs with the single-handed setup.

One hurdle in this approach is the time it takes to perfect a two-handed rod or Spey-style cast. While there is an incredible amount of casting resources and teachers, it still takes time to understand the nuances to become proficient. Most anglers will spend the initial casting time fighting single-handed muscle memory and the desire to push or put a lot of power behind the cast. This is a logical mental process. However, despite the longer rod and heavier line, the opposite is true. When anglers can focus on dialing back the physical power and assertion, they can more effectively manipulate the line and fly.

Another consideration lies within the fish-fighting phase. While the longer lever is ideal for better line placement and fly delivery, it will require some practice when landing a fish. As a wading angler, this is not as detrimental but should be in the mind of solo anglers or those with limited space to maneuver along the bank. It is as important to plan for what is in front of you or your casting target as it is to consider the amount of brush and noise behind you. If you are in thick cover, with only a few feet from trees and brush to water, this style of rod can open a lot more water. However, slow, froggy-type water or small streams are not the place for this tool.

Fly Reels

The floating angler should utilize the spare spool to adjust to changing scenarios. However, mobility is best achieved when the spare spool stays in the truck. While a great asset to keep in the boat bag or packed away nearby, the time and space commitment required to bring spare spools in a pack is often not worth the possible outcome.

Those on a long hike due to a campsite situation and longer than normal fishing time may balk at that suggestion. The few-hours-a-day hiker and wader should strongly consider that approach. While pack horses and trucks may be perfect for slogging the extra spools to camp, a few hours wading and hopping access points should be spent on pre-rigged and ready tools.

When selecting reels, it is wise to be leery of their overall physical weight. Unfortunately, in recent years hot-button phrases like "swing weight" and "balance" have become an all-too-fixated point in fly reel technology. While monitoring weight savings is important, it is not the end all, be all. We as anglers have been brainwashed to think that a perfect weight pairing to a rod is the most crucial factor, that having a perfectly balanced rod and reel will make or break your fishing day. It's simply not true. While in a high-stick or Euro nymph setup every tenth of an ounce counts, those minuscule weights aren't worth splitting hairs with larger tackle.

The same can be said for longer and heavier rods. In the race for the lightest and most refined options,

Overall reel weight should always be a consideration, but a reel that matches the needs of the angler's technique and style is often more valuable in trout fishing. Don't let ounces and drag pressure weigh too heavily in these decisions. PHOTO BY NICHOLAS CONKLIN

Whether classic rods and reels or more modern tackle, tools for fly fishing can be both expensive and sentimental (or just sentimental). Carefully packing and hiding these items in dirty and worn clothes can keep prying eyes and sticky fingers from stealing or damaging your tackle. Outside rod racks and storage devices can be equally helpful to keep from moving rods in and out of homes and vehicles. PHOTO BY NICHOLAS CONKLIN

anglers need to consider the different approaches in rod design versus reel design. In the same way that the most lightweight, thin-walled rod blank isn't always the best tool for all conditions and fish, neither is a machined or cast aluminum reel. The casting technique and position of the arm and rod during the cast are what will either highlight or mitigate the disparity in reel weight.

The theory of swing weight is both flawed and subjective. The overall weight and perceived feel can change due to a variety of factors. The type and size of the line on the spool, the amount of backing, and how much of that fly line is outside of the rod tip will change the balance. Accurate information will not be gleaned by simply holding a fully spooled reel in a shop. All of these feelings can be either exaggerated by poor casting form and stroke or unnoticeable throughout the fishing day.

What anglers can do to better understand the overall weight in the system is to go out in the parking lot and cast the rod. Experiment with different reels to better understand how they truly function with the rod. Compare the different rod and reel pairing options, and solicit advice from the retailer on why they like one pairing over the other. Outside opinions should help form ideas, not lock them in the box of "right or wrong."

The same sentiment can be applied to fly reel drag pressure. While an important attribute in larger tackle and big game, a high-tech, fully sealed drag is just extra noise in the universe of the western trout fishery. There are many fabulous carbon-fiber, Delrin, and truck-stopping drag systems being produced for light fly rods. Because of material advancements, manufacturers have expanded their reach when it comes to refining products for the freshwater crowd, resulting in more time playing with the drag in a fly shop than in a realistic fishing scenario. This is one of the reasons classic, click-pawl reels have come back in style. Outside of their traditional, antiquated (but beloved) look, these reels are simple to maintain and often have a classic-style clicker that, when activated, can heighten the enjoyment of a hooked fish.

When researching fly reels, consider the weight, but put more emphasis on the style and look. A freshwater, trout-specific reel is there to hold the line. Purchase something that has the necessary capacity and style to suit your desired fishing. Overall pounds of pressure and the .005-ounce difference from one reel to another is not something most people will be able to notice.

Don't take it too seriously, and keep it all within reason. This part of the equation, like all the others, should be fun. But here you can play around with the style and color and still be confident you are getting a component that will be up to the task.

The Fly Box

Most well-tenured anglers, either professional or recreational, who have consistent success will typically coordinate flies in the most efficient order. Select a handful of flies, whether it's nymphs, dries, or streamers, and in place of purchasing one in every color and the same size, try buying a few of the same patterns in varying sizes. This approach can be critical for the wading angler, who wants to limit the number of boxes and general tackle carried. Float anglers will have a bit more flexibility here but should still be cognizant of the number of flies. The more to look at in a box, the more distracting it can be.

In my (Nicholas) Euro boxes, for walking and wading, I'll arrange the same style/color of Perdigons in sizes 12 to 20. There may be only two or three distinct color patterns, but the varying sizes can add an extra edge. Different bead or weight sizes in these more high-stick or Euro-specific nymphs can be beneficial as well. Varying bead size gives you a greater range in deeper, more swift pools. It can also give you the flexibility to fish one fly, get into the target zone, and potentially avoid varied currents and flies pulling against each other and killing the natural drift.

The same should be done with streamers. Anyone who has fished a midsize streamer in the West can attest to the value of Woolly Bugger imitations. Bead head, cone head, or unweighted, this pattern is incredibly versatile and attractive. Targeting one or two of the most common color patterns, with an emphasis on size diversity, will allow packability and function.

The wading angler should emphasize smaller, more compact fly arrangements. While you may feel like you are missing out, or not packing enough, keeping your fly boxes light and mobile will pay off. Traveling with a mother-ship box and a few thin,

Organized flies can lead to quicker retying and more active fishing. Orienting by size, color, or preferred patterns within easy reach, based on the dominant hand, will only help to speed up a fly change. PHOTO BY NICHOLAS CONKLIN

Precise fly selection leads to the flexibility to fish confidence flies in various sizes and bead weights. Some anglers even go through the process of labeling bead size inside of a fly box for quicker selection. PHOTO BY OLIVER SUTRO

Mother-ship boxes can allow the wading angler to keep a few confidence flies on hand while having access to a more expansive box, but without having to carry it on every wade. PHOTO BY NICHOLAS CONKLIN

small boxes, you can replace or switch out flies during the day or at the end of it. Most shop employees will sound like a broken record regarding fly patterns, but, regardless of the limited attention span of their customers, patterns are recommended because of success rates. Like a guide, they are trying to give their customers the best possible opportunity, working with a truncated, proven set of tools.

What can set you apart is your ability to fluctuate in size and weight. We're not suggesting going to the discount bin and pulling out the dusty, size 4, double-hook Atlantic salmon shrimp pattern that has been sitting around since time immemorial. What we are suggesting is to be creative in fly selection, take a chance, and pay attention to the bins that are consistently low or out of stock in certain patterns and sizes.

Another way to further create space between you and the average angler is by effectively organizing your flies in a box and labeling them. Small-profile, weighted flies such as high-stick/Euro/Czech/Polish, whatever preference is ascribed to, are very nuanced. Placing them in some ordered system of color, size, flash, body material, or eye orientation will make it that much easier to find what you need

Clean, clear labels, either hand-printed or from a label maker, make pulling the right box much easier. This way you will have the feel and the visual aspects, ensuring the right box every time. PHOTO BY OLIVER SUTRO

An inexpensive cooler or plastic baggie can serve as a handy fly caddy, especially for streamers, where it makes more sense for a wading angler to use a plastic bag with a few standbys rather than tote around a giant streamer box. While popular, branded, expensive, fly-specific storage items seem important, anglers need to lessen the burden on the environment when purchasing limited-use items. PHOTO BY NICHOLAS CONKLIN

when the fishing is hot or daylight is limited. Whatever color, pattern, or hatch you have in your box, the ability to arrange them in a commonsense way and keep every variation in the same lineup will only make you more efficient. This approach requires a commitment to the organization of and replacement of flies.

The thickness, length, and feel of a box can also make your fly selection more efficient. Bury in memory the feel of your nymph or dry-fly box. Whether fishing in low-light conditions or with a hip pack and having to reach around while standing in the middle of a river, being able to quickly identify and pull your desired fly box saves time and can prevent disaster, such as stuff falling out of your pack.

Buy a label maker. It doesn't cost much and can allow for clean and clear box labeling. A marker will do in a pinch, but if you change your mind or switch the fly style, you're screwed. But that's all right—just buy more boxes from your local shop.

Listen to the locals, listen to the shop employees and guides, but don't be afraid to upsize or downsize. Organize and label. Find some consistent box sizes and learn how they feel. Remember, the best fly in the world will not work if there are no fish where you are tossing the line.

Hip Packs, Chest Packs, Fanny Packs, and Storage

The hip pack, chest pack, or fanny pack is a useful tool for the wading angler and one that can be equally frustrating. An appropriately sized, lightweight, and transportable gear carrier, with few but appropriately sized pockets, can be a critical component to success here. Remember, all of your gear will be carried, so, unlike the boat bag, here is a place to trim down and be highly efficient. Unfortunately, with all the options available, packs have become one of those expensive fly-fishing elements that can require trial and error.

Most anglers likely started with the fly-fishing vest. They were clunky, hot, and generally a hassle. While the theory of keeping items in front of you and within easy reach was simple and effective, it didn't always have the intended outcome. Pockets became overloaded and crammed with as much as possible, which effectively diminished the reason behind spreading the weight over the shoulders and back.

These complications helped spur the focus of shifting the weight from the back and shoulders to the core. The industry transitioned to smaller, more waist-focused carrying systems. Encouraged in part by the hiking and backpacking industry, the need to shift weight to the strongest part of the body opened new doors for better tackle management. If workout fads have ingrained anything into our collective psyche, it is that the middle of the body—the core—is one of the strongest areas, and one we all need to focus more on. Even with a dad body, the hips and middle portion of the body can be the most effective when carrying weight. So, the focus continued to shift to waist-style packs that concentrated the load on the center of the body.

As fishing experiences and interests expand, so will the need to adjust how tackle is carried. Some scenarios may even require the use of several packs or bags throughout the season. Suggesting there is a singular option that will suit you from the time you pick up your first fly rod until it's time to join the worms would be unjust. Like all things in sport, this will be a dynamic experience, and being open to change will only afford more opportunities for success.

A few things to consider when selecting a pack or deciding to replace the junk-filled hip pack currently on the roster can be broken down into four "thought" categories: overall size, ease of finding what you need, access points, and waterproofness.

Roomier, better-weight-displacing packs have become common for the western angler fishing in warmer, dry seasons. Anglers who look outside the traditional fly pack or vest may discover something better suited for their needs. PHOTO BY NICHOLAS CONKLIN

OVERALL SIZE

How many fly boxes, reels, spools of tippet, and other items need to be packed for a wade trip? Remember, all of this weight will be strapped to the body and carried all day. Before each wade, reconsider the fly boxes, weights, indicators, and tools that are brought along. Make mental notes during the day and during the season on just how often that tin of size 24 dries or backup shaker of powder is needed.

The length of the walk to the water will also factor into carried gear. If embarking on longer or full-day hikes where additional water or snacks are needed, consider looking into more hiking-specific hip packs. These can be great for anglers who need to carry larger items (or items for kids!) and typically have a more developed strap and weight-displacement

As a wading angler, all boxes and tools need to be thought of as weight on your body. Exploring the size and carrying capacities of various packs and bags can keep what you carry light and highly efficient. PHOTO BY NICHOLAS CONKLIN

benefits. While they often can provide a more comfortable carrying system and larger cubic inches of capacity, most tend to lack the level of water resistance of fishing-specific packs. Consider the most common environments you will face and what exactly you will be packing to differentiate how much water resistance or waterproofness is needed.

EASE OF FINDING WHAT YOU NEED

Too many or hard-to-open pockets on a pack are a hindrance. While companies may upsell the public on the amount of pockets and storage space, keep in mind how easy it is to store something in a pocket. Zip it and forget it. When the season ends, you'll find duplicates of the same thing you barely even touched throughout the year. Keep it simple, easy to find, and easy to open in low-light and difficult scenarios. Zipper size and location will surely haunt an angler if ignored at the time of purchase.

Being unable to locate your tapered leaders, split shot, or extra can koozie because they could be located in multiple pockets is a hindrance. Now just imagine doing it waist-deep in a run, while bugs hatch and fish nip the surface all around you.

ACCESS POINTS

How are the top or side entries oriented? Is there one zipper, a hatch, or is it magnetic? Can you comfortably and safely reach over while standing in the water and open it without much effort?

Do you always need to unhook the pack and sling it forward? Can you easily find what you need, often with one hand? Anytime you consider purchasing a new hip pack, do a simple test. Load in a handful of fly boxes, strap it on, and try to open it without side-to-side movement. You want to be able to stand in the river, reach around, open the pack one-handed, and find what you need.

For those who prefer not to do that or no longer have the dexterity to make that mid-river maneuver, consider a front-facing, or chest-style, pack. These have become a critical tool for serious competition

A dry run of a pack's maneuverability inside a shop can help provide the ability to judge how easy it will be to reach back and manipulate the zippers. This mid-river dance can be tricky, so a pack that fits you, your dexterity, and arm's reach should be the focus. PHOTO BY OLIVER SUTRO

anglers and can compactly contain flies and a spool of tippet with ease. Eliminating the awkward contortions required of those with hip or sling packs can speed up the process of a fly or leader change.

When shopping for packs, spend a considerable amount of time in a shop or other location with a few different options. Fly-fishing tackle shows are an excellent location to play with multiple packs in a crowded foreign environment. Try on several, move around, and twist, turn, and swing them around, if necessary, to generate as realistic an idea of a pack as possible. Selecting a pack, or at least the ideal pack, is not something effectively done through digital resources. While easy to maneuver in the confines of the room-temperate shop, real-world conditions will not be so kind. Can these zippers be opened and closed when the rain is pouring and winds are whipping down the canyon?

Much like a rod, the personal preference factor is sky high, and it is a piece of equipment that will evolve with your fishing. For 99 percent of anglers, it's not a one-and-done.

WATERPROOF OR WATER RESISTANT?

Again, when selecting a pack, consider the material and how and where you intend to fish it. Do you plan to wade deep, or are you fishing in rainy or snowy conditions? Consider some of the roll-top hip packs. These tend to be in the "waterproof" category and have great zippers and a simple internal layout. While there is limited internal storage space, this can be a great option for those who overpack or misplace items.

Keep in mind that packs designed to be more fully waterproof, as opposed to water resistant, can often be more difficult to open. To ensure the watertight seal, a higher grade of materials is used, which can cause trouble opening and closing zippers. There are trade-offs to ensure everything inside the pack will stay dry and mold-free. Anglers need to honestly discuss with their "licensed" shop employee which one is best for them.

Nets

The landing net is a piece of equipment that can further distance the marginally successful angler from the pack. How an angler operates a net and

Nets can be as personal as a pair of waders. Anglers should factor in individual height, distance from the water, and the size of the handle and hoop during the purchase phase. PHOTO BY NICHOLAS CONKLIN

By keeping a low rod angle, the fish is fought off the butt section of the rod, ensuring a proper hook set during the fight. This is also critical in keeping the fight short and preventing exhausting the fish to a dangerous level.
PHOTO BY NICHOLAS CONKLIN

the ability to safely land and release a fish speaks to a more important philosophy needed in modern angling. The obvious benefit is that an angler can ensure a solid hook set and allow themself a selfie with the landed fish. The more important idea is the protection of fish and tackle.

As fish are more pressured, a safe, quick release can do much to help reduce the individual footprint. There likely isn't an angler who hasn't made the offhanded decision to leave the net at home, only to find they've hooked into a large fish in difficult water. While the curse of the net is modern-day gibberish, there is always hesitation when dragging along that last piece of gear.

An appropriately sized landing net is important for two reasons—one immediate, one not so much, but you truly should believe it is a factor. The most obvious is simply for maintaining the health of the fish. The fighting phase, while short on the end of the angler, can cause certain strain and wear on the fish. That problem is only compounded when an angler (even those dipping their hands, with the best intentions) goes to grab a fish and starts jiggling the hook out. The fish flops, and all of those powerful grabs to attempt to subdue it and remove the hook can be detrimental to the fish. While not immediately noticeable, repeated hooking, landing, and aggressive fighting of fish can impact their health over time.

The second part of this equation is the ability to protect the rod. The one-handed fish-grab shuffle dance is not a new TikTok fad, but rather awkward steps that can lead to a break-off or broken rod. Without a net, the rod bends at a candy cane angle, stressing the rod and putting slack into the line. The added slack and diminishment of actual pressure on the fish can lead to a hook pulling out of the mouth.

A broken rod tip means you're done fishing. Be sad, reel in, and go home. Of course, the wise angler has another rod or seven rigged and ready. But the equation still works out the same, and without the aid of a net, it's likely to occur sooner rather than later. And for some anglers, that rod could be out of commission for a season. So, by adding a net to the arsenal, anglers can fight the fish from better angles, with a low tip angle to smoothly maneuver the fish into the hoop.

CHOOSING A NET

Net lengths can vary anywhere from the minimalist length of around 2 feet, with a hoop and net designed for small-water fish, to large, long-handled, hooped nets for big-water trout. These massive nets can be as long as 52 inches and are typically better suited to floating anglers.

The wading angler should consider handle length, hoop size, and physical height when selecting a net. Overall weight and packability should also factor into the purchase. Remember, this net will hang off the body, so will the length and weight impair mobility? Will it get in the way while ducking under trees and hopping boulders?

Many talented custom wooden net builders have arrived on the angling scene in recent years. A long-standing tradition, these modern works of art are impressive, but they aren't always the most portable. The custom inlays and use of various kinds of wood can make for pride pieces, but wood is an antiquated material. It's best to save these nets for the boat when weight and size don't factor in as much. Modern carbon fiber and other synthetic nets can weigh as little as three-fourths of a pound. Those are serious weight savings for the mobile angler.

Another component is water size and your relation to or vertical distance from it. When fishing longer rods or targeting fish from an elevated stance, a longer-handled net is advantageous. While it can take time to become comfortable with the handle length at rest, the extra reach when dipping to scoop a fish higher off the water is helpful. The relation to the water, rod length, and physical body size should factor into your buying decision.

I (Nicholas) stand about 6 feet, 3 inches, with a wingspan closer to 77 inches, so while I have the extra reach, I prefer my wading nets to measure out around 41 inches. While some may question the longer handle when moving between fishing locations, I've never found the length too obtrusive. Anglers should also consider personal body proportions and leg length when selecting a wading net. My

Proper net selection is for more than just landing fish. Proper use of a net will allow for better protection of the fish and the longevity of the resource. PHOTO BY NICHOLAS CONKLIN

Two anglers, two hoop styles. While seemingly similar in width and length of the hoop, the netting will play a role in fish landing. PHOTO BY OLIVER SUTRO

coauthor tops out at about 6 feet and tends to prefer a shorter-handled net, although the first thought normally would be, wouldn't the taller angler want the shorter net?

The longer handle is also an advantage when landing fish for others. It adds a little extra reach and the ability to go deeper into the water column if the fish spots the net coming and tries to evade the scoop.

Again, it comes down to the relative distance from the water and the length of the rod being fished. Additionally, the ability to land and release a trout quickly and safely, while keeping all parties (and equipment) safe, is benefited by the use of a net.

Packing for the Wading Angler

Whatever your mode of transportation, efficiently packing, moving, and unpacking fly-fishing gear can be complicated. Oh, and it all needs to return home in good condition. In chapter 2, we touched on some methods of packing for both the flying and the driving angler. Taking a deeper dive, there are some more advanced tips and critical steps to follow after your trip.

Even those carrying on four-, five-, and even six-piece rods get smirks and admonishing glares from fellow passengers and flight attendants. Where have all the architects with their large, colorful sketch tubes gone? They provided decent cover for an angler to slyly move unnoticed past the gate guards and into their seats. The angler with the luxury and proximity to drivable western fishing locations may seem to have the upper hand, but always be wary of the overload. Just because it can be packed doesn't mean it should.

Airline-hopping anglers, although confined to limited cubic inches, need to prepare equally diligently to avoid forgotten items. Believing that the fly shop in town will have anything they may have forgotten, they can find themselves in a crisis when they hit the local shop only to find out every other angler in town is scrounging for that one special fly or pair of those new rubber-handled forceps. Don't leave to chance what proper diligence can alleviate.

Wet boots and waders are no fun and need to be cleaned and dried in case of travel delays once packed away. Most anglers fail to realize just how much time these items spend in the water. PHOTO BY NICHOLAS CONKLIN

One proposed strategy to ensure gear arrives and is easy to locate is the separation of church and state. By separating portions of tackle between a checked bag and a carry-on, anglers can prepare for any eventuality and still have the basics in the event of lost or delayed luggage. Another option can be to mail larger, heavier pieces ahead to a familiar and trusted location. This approach can provide much smoother on-the-ground movements if hopping between cities and multiple planes.

Odds are that gear like waders, jackets, and packs will be subjected to wet weather and submersion as a walking angler. If the tendency is to quickly stash away wet waders and other gear in the dark recesses of travel bags, then stinky, moldy gear will be unpacked once home. If travel delays are encountered, flights are missed, or bags are lost, a sour surprise is sure to appear on receiving baggage.

During an especially cold and rainy trip to British Columbia, I (Nicholas) failed to take an attentive role when repacking, and I paid for it when the bags arrived over a week later. I ignored my own rules and tricks to help the gear dry out during travel and was greeted with moldy, sandy waders and boots. What was originally going to be a simple day flying back to the States turned into my bags being stranded in Alaska. Eventually, they wound their way down the West Coast, having made some further weather-delayed stops, before arriving in Dallas, Texas.

After many hours in scalding hot water and a spray bottle of cleaner, I was finally able to de-mold and de-stink my gear and return it to operable condition. Unfortunately, the spots inside my waders never really went away. Nor did the mildewy olfactory persistence of the attempted post-trip cleansing ever really abate.

What this experience forced me to do was more deeply analyze the pre-trip and post-trip packing to ensure the cleanliness of my gear. One straightforward way was hand towels. Now this idea may have come about from many nights in hotels and the variety of small hand towels and other nondescript white cotton towels. Without naming brand names, or outing myself as the masked towel thief, I would occasionally slip one into a pair of boots or take a larger towel and wrap wet waders before placing them in my baggage.

Despite bordering on a misdemeanor, it not only helped keep the rest of my luggage dry and grit/sand free but also aided in absorbing the moisture that clung to my wading apparel. Smaller towels roll up perfectly into wading boots and shoes and are always handy to simply set wet gear on at the end of the day. These smaller towels are ideal for cleaning up fly lines and backing at the end of the trip. While it may seem that the high-priced, new-age fly line does wonders in repelling water, it also traps that water, burying it into the line and spool. This situation leads to mold and noxious smells. A minor annoyance, if left unclean and undried, it can lead to damaged line coating.

Some preventative maintenance of drying out the length of the line will only add more days to it. Of course, if coming off a particularly muddy, wet, or otherwise weather-offensive trip, fully cleaning and drying the fly line is recommended. A hotel sink, hot water, and a bar of soap can do wonders and keep the potential for mold and line damage at bay until the return home.

Of course, maturity and salary have moved me to the actual purchase of towels and traveling with them, but the early days demonstrated how a simple measure can save time and money in future tackle replacement. Outside of the moisture absorption, wrapping pricier gear in cotton towels can keep the sticky-fingered baggage handlers at bay. Similar to lining the top of your baggage contents with your worn underwear, wrapping your high-end waders and/or wading boots in a simple, dirty towel will often help dissuade people from pulling out and misplacing these critical items. No one wants to poke around something wrapped in a wet, dirty towel, just like no one wants to shift around a few pairs of week-old underwear. It's a strong deterrent.

Silica or desiccant packets can also help rid your luggage and tackle bags of unwanted moisture. Fresh packets bought solely for this purpose can aid in moisture reduction. Our apologies to those out there looking to save that packet from the beef jerky bag and repurpose it. The appropriate silica packets can be bought in copious quantities, in assorted sizes. Charcoal bags have also been proven useful for slipping into checked baggage and anywhere wet tackle may be sitting in a dark, damp environment. Smaller ones can work wonders in fly boxes, where large, feathery streamers and other natural-material flies can stay wet for several days.

These items can be especially helpful if boxing up and shipping waders, boots, wading packs, and so forth. As the shipping of items may involve extra road time, some cleverly placed moisture-removal items can make unboxing much easier. It is advised to confirm with the shipping company what items can and cannot be shipped. Our experiences indicate that most desiccant-type materials are designed for travel and pose no shipping concern.

Various rods, reels, and lines can and should be utilized across the angling universe. The walking angler has many of the same tools that will work effectively in a boat. However, with some subtle tweaks, different approaches will help the floating angler achieve just a little bit more success. PHOTO BY OLIVER SUTRO

CHAPTER 8

TACKLE FOR THE FLOATING ANGLER

Tackle is a personal choice, and we exist in a world of constant innovation. Like the materials used, tackle choices are endless. In the coming pages, we carefully dissect the commonly used and/or available rods, reels, and flies and why we chose the gear we fish. Whether a floating or a walking angler, many of the same tool selection and organization ideas are transferable. However, the floating angler should realize that just because they aren't lugging this gear on their backs, it's not an excuse to load up on a rod and reel for every possible scenario.

Much of what can be used wading can at times be prohibitive when in a raft or boat. These close quarters, often with one or more people, can create a bit of chaos.

Despite the large numbers of drift boats, rafts, and other vessels purchased during the COVID-19 buying spree, we still want to address how to prepare for fishing out of other anglers' boats. While the allure of having your own boat is high, there is value in utilizing the kindness and services of your friends and local guides. Stress and anxiety will be reduced and enjoyment increases as a rider rather than an owner. There are some great books dedicated to owning and operating a drift boat or raft. For specifics on maneuvering and safe operation, we recommend Neale Streeks's *Float-Fishing Strategies*.

Texts like these are great skill-building for the new or intermediate boat owner. We will not be discussing the ins and outs of rowing, but rather how to approach the tackle component of owning or fishing out of these vessels. These are a few suggested approaches for boat owners and non-owners alike.

Choosing Your Rods

Most drift boats and rafts have storage for four assembled rods. If you consider that likely two rods are being used simultaneously, that means a boat can safely manage six assembled rods. The authors spend much of their freshwater angling time in the Madison Valley. On any given day, these rivers and creeks can be a tight-line nymphing dream, the best streamer river in the Lower 48, technical dry-fly water, or all about terrestrial hopper-dropper-style rigs. This can present an opportunity for "too many" rods.

Whether in a personal boat or going out with a guide, the key here is communication. Like many of the concepts in this book, a clear and verbalized plan can make for a successful day. While all the above-mentioned tactics are possibilities on a river, the sharp angler will understand their time and efforts are best spent committed to one or two methods.

Whether fishing with a paid guide or not, having more rods (which is more clutter) than necessary is not a welcome addition to a day's float. If you want to fish a specific way, say so, but ensure that you are properly equipped and have realistic expectations. Too often the desire to fish a certain technique and put forth effort assumes success.

As a floating angler sharing a boat, having a few rigged rods can aid in success. Available space and the most effective use of that space should always be considered. This issue can be an equal inconvenience if carrying one rod and constantly changing flies. The solution is not always carrying fewer rods.

Here are some thoughts on what and how to carry multiple rigs when floating in another angler's boat. Depending on flow, bug activity, weather patterns, and seasons, all of the below can provide the critical link to improved success. But this is an area where we hope to impart the desire to be as technique-focused as possible when fishing out of a boat. While the owner and main rower of the boat will be able to slide several different tools into the holders, the co-angler likely won't have that freedom. So, the initial focus should be on each boat occupant's ability to pull and replace rigged rods while moving. Simply understanding the choreographed dance that occurs between a rower and anglers during casting and fish fighting is paramount. The second focus is understanding the likelihood of success, as well as what that success truly means.

On one trip floating the Gunnison, there was a co-angler who only wanted to throw dries in front and streamers in the back. Three fish were caught that day, which happened to represent the largest rainbow and brown of the season for that boat. Another boat encountered that day had focused its efforts on tight-line nymphing. After several hours of this, we acknowledged the other boat's many smaller fish landed. They did the same for the vastly fewer but much larger fish from our boat. As the day ended,

Before loading rods and hopping into the seat, anglers should always consider tool placement and the ease of access while on the water. Don't let the ease of access and ability to carry a lot of gear make life crowded on a boat. PHOTO BY OLIVER SUTRO

we met up again at the takeout and overheard one angler questioning the guide on why they "didn't catch more fish like the other boat's."

Their expectation was not based on reality or communication. They were likely handed an indicator-style rig and told to fish it all day because of the likelihood of success, given their experience and skills. If you intend to streamer fish during the float, stick to a streamer rod. If you want to commit to a tight-line approach and develop those skills, bring that specific rod (or two) and commit to fishing the technique throughout the day. The ability to solely focus on one style will do wonders to improve your overall game. If those skills and wishes aren't communicated before a float, the guide or boat owner should and will do what they think is best for a successful day.

THE HOPPER-DROPPER, DOUBLE-DRY SETUP

Likely the most common setup, the standard 9-foot 5-weight will be in every angler's quiver from New Mexico to northern Idaho and back. A 6-weight rod can allow for easier delivery of larger flies, and it also is a more well-rounded tool that can handle some of the light streamer applications. In most scenarios and most seasons in the West, this type of setup will be the most versatile.

Whether the angler utilizes a standard 9-foot 5-weight rod or graduates to a 6-weight, the ability to fish these multi-fly rigs will be powerful. While both weights are universal and can work in many fishing environments, anglers should also consider the length options. As discussed in chapter 5, the floating angler needs to consider their physical placement differently than that of a wading angler. While the wader can have a bit more control over body position, the drifter is often at the mercy of the rower's aptitude.

Beyond the rower's abilities, river dynamics and obstacles may also put the caster at either an advantage or a disadvantage. The bias toward the angler in the front of the boat will typically put them in a shorter casting range. Assuming anglers are floating with the intention of fishing along the banks, the angler in the front will likely have a preferred position. However, with the addition of a longer rod (9½ to 10 feet), the angler behind the rower can diminish that casting distance advantage. The longer rod lever will also aid in efficiently picking up and repositioning lines and flies. As the guest angler, the simple rod spread can be switched as fishing positions change throughout the day.

This is a wise idea for boat owners as well. They can pack along a tool that can fish multiple techniques but also grants a dedicated location in the boat. The front angler has a shorter rod (9 feet), while the caster in the back has a longer rod. While the anglers may move, the tools stay put and ensure their roster position. The alignment or position of the rods and tackle, relative to the limited movement needed to use them, is key to a clean boat.

HIGH-STICK/EURO NYMPHING

The high-stick, or European, style of tight-line nymphing is an effective tool for the wading angler, opening huge portions of new and productive water. For the floating angler, this technique tends to skew toward heavier rods and heavier flies to effectively get the offering into the zone.

For boat-bound anglers, we typically don't advise the longer, lighter 2- and 3-weight rods. While helpful when wading, these rods present some challenges to the confined boater. The first is length and maneuverability. With rods upward of 11 feet, the simple task of storing in a boat is hampered.

Big foamy flies need a little more mass from the line, requiring a bigger rod. The heavier rod will be a benefit when big western winds whip up.
PHOTO BY CAPT. COLLINS D. ILLICH

While a great tool for tight-lining some of the larger, heavily weighted rubber legs–style patterns, we feel the inclusion of this rod isn't warranted on a float trip. Heavier and longer rods, which can be multipurposed, can keep the overall gear to a minimum and the focus precise. For floaters that utilize a 9- and 10-foot rod system, typically the 10-foot rod can be used just as effectively in this situation. Leaders will need to change, but the length here is effective without adding another rod to the boat.

STREAMER STICK

Whether number one or two on the boat angler's list of rods, the streamer rod has cemented itself in modern history for most anglers. Just after the 5-weight, "do it all" style of rod, most anglers go on the hunt for a meat-throwing stick.

Big, fluffy, and just plain fun to fish, these meaty flies have been an incredibly compelling evolution within the sport. Talented tiers, anglers, and marketing teams have found more and more ways to sell these creatures, with great success. If you want to entice a large predator, you typically need something big. Because of advancements in rod and line technology, the ability to deliver these wind-resistant, bulky patterns has gotten easier over the years. With increased grainage in fly lines and more powerful fly rods, anglers have found that it is much easier to cast big flies on lighter rods.

The 9-foot 7-weight with a full-wells grip and fighting butt is a logical addition. Appropriate for single or articulated flies from 3 to 6 inches on either a full-floating or a sink tip, this rod weight and an appropriately paired line can more than deliver the fly to the target. For boaters with a committed 5-weight, the addition of a 7-weight will provide great technique coverage and two easily transportable tools.

Whether the owner of the boat or just a companion, this system also allows for a bit more latitude when it comes to reels and spare spools. While not unfeasible for the walking angler, simply storing a spare spool, with a multi-density line, or a small wallet of spare tips can be easily done on a boat. Instead of packing along multiple rods, rigged from floating to a full sink, consider the conditions and how a simple tip can be looped to the fly line. The ability to add to your floating line can change your fly position at depth.

Most companies offer tip "kits" that can range in length from 5 to 10 feet. For most western trout applications, we tend to rig the 5- or 8-weight versions. Seldom have we encountered scenarios in which a full-sinking line is needed. However, that doesn't mean there isn't a potential application for it. Like all things here, consider what will be used most and what your focus will be for much of the angling day. Not every eventuality needs to be packed for and stored on the boat.

One item for all streamer anglers to consider is the addition of a second or trailing fly. As the streamer bite in the West picks up and the pressure is on to pack on calories before winter, most anglers will present a single fly consistently along the bank. While effective, this can turn off some fish, as they have likely seen this presentation repeatedly. One way to separate yourself from the field of fur flingers is to add another fly to the rig. A larger streamer with a nymph trailing behind can provide a second point of attraction. More important, it presents the look of a dislodged bug. Perhaps that streamer rubbed the bottom or bumped a rock in a way that dislodged a nymph from its hold. A trout that may not want to chase the fleeing baitfish (fly) now has a simple meal that it can more comfortably move up to and strike.

TROUT/MICRO SWING ROD

This is another tool for the floating angler, which we typically bump down the line of importance. As discussed in the previous chapter, trout/micro swing rods can afford the wading angler a great angle of attack. The longer levers do, however, present some challenges to the boater.

The first is the overall length. This point impacts both the storage and the fishability of a boat. While it can be done, fishing the longer two-handed rod and landing fish safely in a boat is a challenge. This method usually is effective only if the boat is being used solely for transportation and anglers are doing their fishing while standing in the water. When it can be done, it will require some serious teamwork, and typically one less angler in the boat.

The second hiccup is the packability of the rods. Due to their length and the amount of storage space within a river boat, the rods often must be broken into two sections and stored. It generally is more of a hassle than it is worth.

Pairing Reels with Rods

Reels for the boat-bound angler do not need to differ very much from any common wade-fishing setup. When researching fly reels, consider the weight, but put more emphasis on the style and look. Simple material construction, whether fully anodized aluminum or cast aluminum, is more than sufficient in a boat or raft. Many modern reels are both highly durable and cost-effective, an important point to consider when setting budgets and goals for assembling your boat-fishing tackle.

Beyond the color and porting style of the reel, components like the reel handle and drag knob are important to consider before purchase. Larger-than-normal reel handles and drag knobs can be valuable to those fishing in colder, nastier climates or those who may not have the dexterity they once did. While you shouldn't anticipate blistering runs and screaming drags from western fish, an easy to handle and maneuver reel handle is important. As with your wading reel (and they likely should be the same), less focus on drag pressure and the marketing buzzwords will serve you well.

Finding a quality reel that suits your style, has the components you prefer, and fits within your budget is most important. Consider looking into reel brands that offer spare spools should you find yourself in a situation where a line of a different density is needed. As flows or river depths fluctuate during the day, the extra density can get flies to more appropriate depths. For those fishing both fresh and salt water, this is a great, cost-effective way to take your saltwater reel and repurpose it with a spare spool and freshwater-specific fly line.

When seeking out a freshwater fly reel, understanding the application, your budget, and overall use the reel will get is important. Both cast aluminum and fully machined aluminum reels will suffice. Anglers should not fixate on fully sealed drags and heavy-duty internal components, focusing instead on weight, style, and price point. PHOTO BY NICHOLAS CONKLIN

The Fly Box

If you are fishing with a guide, the expectation is they will provide all the terminal tackle (flies, leader, split shot, strike indicators, etc.), so there is little responsibility on your end. You will want to confirm who is responsible for terminal tackle and articulate whether you intend to bring some of your own bugs. Anglers should consider this an opportunity to learn, not just be handed something without asking why.

One of the best, often overlooked aspects of fishing with a knowledgeable guide is questioning why they chose this fly over that and why they are telling a client to fish a fly a particular way. When seeking to build and refine your angling toolbox, look to the professionals and pick their brains relative to fly selection. We also like to use this time to ask why and how they organize their flies. While it looks cool to have a few random, torn-up flies sticking to the weather-worn foam fly pad, there is a real reason a guide sticks a pattern there. As discussed in the wading angler chapter, when organizing your fly boxes, always consider the most used patterns and the ease with which they can be accessed. As difficult as it can be to maneuver a fly box and a rod while wading, consider standing on the bow (or, worse, rowing) and how difficult it may be to retie.

As covered in other chapters, the sport of fly fishing unfortunately requires a lot of trial and error relative to gear selection. Flies are no exception, and neither are fly boxes. While the smaller pieces may not seem as daunting as a rod or reel purchase, flies add up, and several dozen patterns without a proper home or purpose are just wasted money. While there are no true corners to cut to achieve success, an easier road map to success can be made by asking these questions.

Don't wait until you are 200 yards from the launch to realize you have only a few ratty old flies stuck to the fly patch in the boat. An open dialogue about who has what and ensuring it is loaded is key before launching.
PHOTO BY CAPT. COLLINS D. ILLICH

A foam roll with a thicker diameter, like those used by yoga enthusiasts, is great for storing, pre-tied multi-fly rigs. While commercially available containers and specialty boxes exist, thinking outside of the box can provide a simple and effective solution. PHOTO BY OLIVER SUTRO

If you are fishing out of a friend's boat, the onus shifts. This again is an area where communication before the trip is key. Waiting until the morning of (or, worse, 20 minutes into) the float to confirm who has what flies and how you plan to fish can be disastrous. Of course, if you intend to continue your friendship, you may want to cover how you plan to replace or pay for those flies that got hung up in trees and submerged logs.

As a boat owner, being organized and efficient with your terminal tackle is the only way to protect your investment. This ultimately transitions to an enjoyable day. Most gear is kept in a boat bag that rides next to the rower's seat. The bag is loaded and organized, and each box is labeled and color-coded based on the patterns and season. We tend to veer toward keeping the same brand and style of boxes in a boat bag. Beyond brand loyalty, these boxes are consistently sized, ensuring proper storage. This also eliminates any guesswork on how many specific fly patterns will fit into each box. Much time is wasted fumbling around odd-size fly boxes and fly tins to find a certain pattern. If anything, having a box with more space will make finding and storing flies easier. This allows for easy pulling and replacement of specific, trusted patterns dependent on needs.

Those who like to have pre-tied rigs or a handful of flies within reach should consider a foam pad or roll. We utilize a 12-inch-long yoga roller to store pre-tied hopper-dropper rigs. A separate foam roll is also on the boat to accommodate tight-line rigs. So, when fishing is hot and there is no time to properly retie a multi-nymph rig, this organized roll requires only one knot, and it's back to fishing. Pre-rigging tandem flies save time on the water. While commercial options are available, this inexpensive, thicker-diameter foam roll allows for clean access and storage of multi-fly rigs. It also helps mitigate rushed, poorly tied knots.

Efficiency is not just a concept—it means more fish. The last time we floated, we averaged one cast every 45 seconds. Over the course of an eight-hour day, that is 28,800 seconds on the boat, or 640 legit casts. Spending five extra minutes every time you rerig looking for this or that is 53 fewer casts per day—53 fewer chances to catch your fish. It adds up. If you have four days on the water, that is 212 casts or one-third of a day less than your better-prepared friend.

Hip Packs, Chest Packs, Boat Bags, and Storage

When it comes to packs and bags, always be aware of your footprint. Nothing makes a boat less enjoyable to fish than a crowded space. As covered in the previous chapter, anglers should consider four things when selecting tackle-specific boat bags: overall size, storage space, ease of access, and weatherproofness.

Size is the main focus here, as you likely need to store these items on either side of your seat. If the boat or raft is larger or has more convenient options, great, but it is always safe to err on the side of less is more. The airplane carry-on and one-personal-bag mantra needs to be adhered to in this scenario.

When selecting a boat bag, it is important to pencil out the length, width, and depth to accommodate your fly boxes. Trying to cram too large of a box or, worse, many smaller boxes of varied sizes into a bag will only create confusion. If floating with a guide or friend, don't be shy to ask them what (if anything) you need to bring. Confirm that you have a bag and ensure it will fit within easy reach. If relatively new to the boat scene, the on-the-water time can be a great chance to ask the guide why they carry the bags they do. Given their need to be overly prepared for any scenario, the storage devices on a guided boat will be highly precise and focused.

Storage space can be broken down into the open-style bags and those with dividers and multiple pockets. We tend to utilize the more open boxes for ease of viewing and access. Like the wade angler who crams various spools of tippet, flies, and accessories

The angler's boat bag should be clean and packed with enough gear to safely get the angler down the river and back to the launch. A GPS provides more than just a lifeline to rescue. It can also effectively mark takeout points and on which side of the river the trailer will be parked. PHOTO BY CAPT. COLLINS D. ILLICH

into any available pocket, a boat bag with hidden pockets can lead to lost or forgotten items. We find it's best to keep things in the open and easy to find. The larger zippers enhance grip in foul conditions or when wearing gloves. When a fly fisher has poor manual dexterity or foul weather conditions affect maneuverability, these larger, higher-quality zippers are significantly easier to use.

Access points are key, as the nimbleness required to navigate a boat bag while cruising down the river can be tricky. We look to bags with roll tops or larger-than-average zippers when gearing up for the day. The ability to quickly access and pull fly boxes, leaders, and other tools, often one-handed, can be valuable.

Weather resistance is the final step in selection. Various materials can provide different levels of waterproofness and moisture protection. What it comes down to is the price. As anglers, we are often watching the skies to provide us with information. This should be no different when standing in the confines of a comfy fly shop or staring at a shop's website online. How and when will you be using this bag, and how wet do you think it really will get? If this is a bag that will spend most of its life in the boat, out in the elements, a deeper financial commitment should be made. If it's a simple go-bag that lives in the back of the truck and gets pulled out when that hopper bite is going off, a more cost-effective approach can be taken.

The type of water should also be examined, as the nature of the river will provide either a relatively dry or a watery environment. If most floating is to be done on calm, gentle water on the Missouri River, for example, less emphasis should be placed on a fully waterproof bag. If floating in rafts, in whitewater scenarios like the Blackfoot, then something that can be submerged and stay dry is necessary.

For those who want to bring along a high-end DSLR camera or other weather-sensitive gear, waterproof sacks are often needed. While a bag may tout its "waterproofness," it is often better to be safe than sorry. A simple plastic ziplock bag is great for mobile phones and wallets, but anglers here should consider something more substantial. These waterproof sacks and cube-style bags are also beneficial for anyone packing a spare rain jacket, hat, or gloves. There is truly no greater annoyance than the sleeves or hood of a rain jacket popping out of a boat bag. This impedes progress when selecting a new box and often leads to the jacket spreading and taking up more space than needed. Always consider these extra items and how best to contain them.

A captain's boat bag or bags will typically contain more items for the day. These can include medical supplies, spare boat plugs, sunscreen, and bug spray. While both the captain and the anglers should lean toward overpreparedness, this is an important conversation to have when heading out with a guide or new fishing companion. Those little things often overlooked can affect the enjoyment of the day.

There are endless storage opportunities for your boat tackle. Simplicity should outweigh advertised features.

Nets

On a boat, you need only one net. For those with a boat or considering outfitting a raft for float fishing, the net is another piece of gear to view with the totality of the boat's space in mind. This thought should extend beyond the physical size of the net to the ability to store, pull out, and replace it as easily as

One lightweight, long-handled net is more than sufficient for easy maneuverability around a drift boat.
PHOTO BY CAPT. COLLINS D. ILLICH

Larger, longer nets are ideal when fish need to be landed outside of the boat, often in challenging water.
PHOTO BY OLIVER SUTRO

possible. A boat net with a deep hoop, wide basket, and long handle is key.

A handle around 4 feet in length is a strong starting point. This will function well in a pontoon-style raft as well as a high-sided drift boat. The handle should be just long enough to easily maneuver over the gunnel of the boat but short enough that it can be managed with one hand. While the wading angler needs to compensate for the net's length while carried, the rower must consider the space and crowded excitement when a fish is hooked. Seldom will the boat and rower be in an ideal line when a fish is hooked, so the ability to maneuver and safely land the fish while keeping the boat's occupants safe is necessary.

Longer-handled nets can also provide an element of safety when a fish is hooked outside of the boat. Acting as a makeshift wading staff, a longer handle can help the angler and the net person safely manage fast, rough water.

Thoughts on Fishing with a Guide

Usually, there should be a line of communication the night (preferably several nights) before a trip with a professional guide. The communication is likely to focus on the "where" and "when" to meet up and what to expect. What is often missed here are the hard questions and answers. This is the time to set the standards.

While it's exciting to think of the early morning launch and what's going to be served for lunch, more direct conversations should be had. It will give your guide a legitimate sense that you are committed and want to learn. It will also create an understanding of your angling skills and objectives. The number one suggestion is to listen. This is the time to listen to understand, not just reply. Guided trips can be costly, but think of them as the CliffsNotes version

Having a clear game plan on expectations and an honest assessment of potential success is vital. This chatter needs to continue as the day progresses and a level of comfort is established. PHOTO BY OLIVER SUTRO

of the guide's skills, experiences, and expertise. They already have a plan that is success-driven for you. This concept should go both ways and is an important step in the development of a relationship with your guide.

If you have a specific concept, technique, or stretch you want to float, this is the time to state those desires. Initial plans and techniques should have been discussed and understood well before the trip. This will make the initial selection and booking of the trip as positive as possible.

We can't tell you how many times we've been party to the awkward, albeit kind responses from a guide just meeting their clients for the first time. While the most successful tactic for the day may be approach A, we've heard the excited, unfocused questions from the sports saying they heard how tactic B was really working! Oh, the poor guide. While doing their best to attempt a polite response, they know that whatever "hot fly" these clients overheard on the plane into town or at the bar the night before is beyond their skill level.

If you only want to refine your dry-fly fishing, even during periods of non-hatch activity, explain that. If you desperately want to float a certain stretch of river so you can drift by your ex-partner's house and look cool, say so. Just like the guide will be honest about the conditions of the river and the bug activity, anglers should be honest about what they can and want to do during the day.

On a Southern Rockies river, I (Collins) once had a client who was not catching anything. As the guide, I had a burning desire to at least hook into a small whitefish, to claim some minor victory, but that feeling was severely one-sided. As the perceived pressure mounted and the day was wearing down, the client and I finally communicated.

"I don't like catching fish. I just like to cast," the angler finally said.

On hearing that, the fly was removed, and the rest of the trip was spent working on different casts and how to cast different rods for different techniques.

Fishing with a Boat-Owning Friend

Your authors spend a lot of time in this category. Like fishing with a guide, your game plan is critical and preferably discussed well before the trip. But this friendliness should not override the need to communicate. This is a situation where you need to be honest about your skills for both success and safety.

It is also the least stressful environment, so there should be freedom for both parties to be totally honest. At this point, you should understand each other well, know each other's equipment, and have likely spent more than a few days on the water together.

Regardless of how or why you fish, the ability to back a trailer or help recover a boat is a plus. While it is advised to stay clear of the launch and load process when with a hired hand, fishing with a friend is different. Whether your thousandth float together or meeting for the first time, sharing responsibilities makes for more enjoyable days. While the obvious conversations on who will bring the cooler of beer and who has extra flies are important, a dialogue about this process is more helpful. Whether launching at a frequent spot or a wholly new location, discussing who does what and what is the best way to load and get into the boat sets the stage for the day. This non-fishing footwork should be the first of many opportunities to build a strong angling team.

Thoughts for the Private Boat Owner

Being a boat owner represents the highest level of responsibility. First, you are responsible for the safety of everyone on your vessel and the safety of those around you. That is before we even start fishing.

While most anglers may jump into the bow seat under-equipped and under-prepared with rods and tackle, being unprepared with foul-weather gear is not as easily compensated for. It can become scary and potentially life-threatening when the weather turns foul and it rains or snows, but the same safety challenges can occur on warm, high-sun days. Most western river floats are not very easy to

While obstacles like a low-hanging bridge are usually apparent, there is a responsibility between the angler on the bow and the rower. The forward-facing angler should always be another set of eyes and ears for the person on the sticks. A friendly relationship should not intimidate someone from mentioning a possible rock or logjam down the river.

PHOTO BY CAPT. COLLINS D. ILLICH

abort. Whether in remote locations or even floating through civilization, it's usually not easy to just pull over and get someone to safety. At last check, we haven't ever floated by a gas station or convenience store along the water.

Hypothermia and exposure are real concerns in the West, even in pleasant conditions. Basic safety items like jackets, hats, and sun protection are a necessity for every boater. This is not an area to assume your co-anglers will come prepared, as it's often a topic that gets lost in the pre-trip conversations.

Every boat should have spare rain gear, appropriate medications, sun and bug care, licenses, and a GPS/location device beyond a cell phone. Unlike the fly gear "go-bags," boaters have the ability and responsibility to always keep these in the boat. While medications and first-aid kits will need to be checked yearly for expiration, clothing-type items should have a permanent home in a seat storage or individual bag.

Boat owners should also confirm float plan details with a third party. While the shuttle driver may know to drop your rig off at a particular location, it is wise to provide a loose outline to someone not on the trip. While you may be tempted to head off for a float without telling your partner exactly when or where you get off the water (in case you want some extra post-trip beers), it means no one will know where you are. It may seem like a flippant omission of details at the time, but providing another individual with a time- and location-specific float plan can potentially save a life. Being overly safe and smart when informing others of your plans and what to do if you don't make it to the takeout at the end of the day provides a safety net. It's never safe, even in this Big Brother day and age, to assume you will have cell phone service or that you will luckily encounter a helpful citizen with time to burn at the boat ramp.

Anglers should not always assume that cell phone reception in the West is a given. A clear plan before the float, such as launch location, planned float stretch, and anticipated time when you will get off the water, is critical communication to have with someone not fishing. Even if floating a familiar stretch, a simple text to a friend about your plan can be a life-saving measure. PHOTO BY OLIVER SUTRO

Don't let the back of your truck look this way. This mess may seem fine now, but there is a tendency to let these masses of gear and tackle linger after returning home. PHOTO BY CAPT. COLLINS D. ILLICH

CHAPTER 9

TRIP ANALYSIS: ARE YOU AS EFFICIENT AS YOU SHOULD BE?

The week is over, all the fish have been caught, and the BS, laughter, and stories have all been told and rehashed. It's time to pack away the dirty underwear, stash wet boots and damp waders, and have a beer before flying out or packing up the truck and driving home.

While this is often the time to sit back and ruminate on the memories recorded and some of the fish landed and lost, it is also the time for the enterprising angler to take serious stock of the trip. By committing to note-taking on a phone or pad of paper, anglers can take what is still fresh in their minds and put it down on paper (or what we consider the modern-day equivalent of paper, the iPhone Notes app).

This is the time to seriously reflect on the successes and struggles during the trip. Memories are fresh, and the mind has a better chance to accurately describe what worked, what did not work, and some ideas to improve for the next trip. Mentally walk through the rods, reels, tackle, and knots tied, critiquing what worked and what did not.

Anglers who truly wish to rise above and become just a small percentage more effective than the next angler always take a deep dive into what they did and didn't do during a trip. While some of these ideas may be bordering on "work," know that by better understanding your approach and continuously working on ways to improve, your overall enjoyment will increase. A more streamlined, tight approach will typically mean more fish. If fish counts aren't the objective, at least a more efficient system will hopefully cut out the unwanted or frustrating components of your fishing.

While inebriated airport lounge conversations and grandiose ideas, shouted over the hum of the plane, all sound good in the moment, there is a sobering period when you deboard, grab your bags, and head home. A lot can be lost in the post-trip adrenaline flush. Don't wait until the next morning or when the facts start to wear away as more poetic memories and experiences flood your brain in the days and weeks following the trip.

Packing Before and After

If your pre-trip packing resembled the walk-in freezer at a Michelin-star restaurant but opening the bags after returning home resembles the piss ditch at Burning Man, you may have a problem. Many hours, days, and even weeks go into the successful angler's pre-trip packing and planning. Less finesse and typically damp, muddy, and grungy clothes get haphazardly tossed into the bag for the journey home. While all too commonplace, anglers need to be wary of those scenarios spilling into future trips. The rushed or last-minute pack jobs can cause misplaced items and lost angling time.

HOW TO PACK TACKLE FOR AIRLINE TRAVEL

Developing a game plan and following that strategy to pack and pull in advance of the trip can ease the last-minute teeth clench on setting the bag on the scale. We've found that packing and assembling

Culling excessive tackle allows for lighter and organized baggage. While most anglers won't bring the proverbial "kitchen sink," like the rack of boots pictured, there is often a writhing fear of forgetting something. PHOTO BY NICHOLAS CONKLIN

gear early and performing the daily unpack can help streamline tackle. Pack all you think you may need, and then some, and spend the next few days pulling out items and deeply evaluating what you crammed into the bag.

Assembling gear early and performing the daily unpack can help streamline what goes on the trip. This pack-and-pull method can eliminate duplicate and triplicate pairs of pants, sweaters, and cotton T-shirts that likely never get worn. This idea should be no different from how you cull rods, reels, and tackle storage items.

Start with your luggage. How do you plan to utilize your baggage? Is that big duffle bag or suitcase solely for clothing? Or will it house your waders, boots, reels, and fly boxes? Can the carry-on bag be repurposed as a boat bag? The ability to multipurpose your traveling bags into your functional tackle-carrying tools can help streamline the amount of "junk."

DISGUISING EXPENSIVE TACKLE

As anglers, many of us have a soft spot for antique and well-worn tackle. The romance of classic reels, well-used rods, and beautifully tied flies adds to the mystique of the sport. While these pieces hold special memories and meanings in our lives, they can be a flashing red light for a baggage handler with sticky fingers.

Depending on your destination, certain types of tackle, and, more important, the "look" of the traveler, can put a target on your back. While not as common in the West, there are undoubtedly towns that slurp up more tackle than others. When flying, photograph all the tackle beforehand and make notes on what was brought and its condition. As mentioned in chapter 7, use the tools, albeit stinky, dirty ones, available to help disguise tackle when going through unfamiliar airports. Worn clothes and undergarments can provide an ideal buffer from someone pulling a classic fly box or early 1900s Hardy reel from your luggage.

HOW TO PACK TACKLE FOR ROAD TRAVEL

For road travel, the Clampett family look can be avoided by the use of large plastic storage tubs. As space allows, transitioning from soft-sided bags to hard rubber tub containers allows for secure and dry transport of tackle. Beyond helping on the road, on returning home, the clearly labeled boxes go right back on the shelf until the next trip.

Available in most hardware and big-box stores and ranging from 7 to 71 gallons, the heavier-grade storage tubs, if not already used for Christmas knickknacks and household tchotchkes, are great gear storage. The 27-gallon tubs, or thereabouts, are typically the best for space and overall weight capacity. Adjustments may need to be made for the size of fly boxes, boots, and so on, but these seem to be the most effective and manageable size.

Outside of carrying capacity, relatively low cost, and ease of transport, these containers are easy to clean. If they get wet or muddy when riding around in the truck bed, or even in a boat, they can simply be hosed off. If you run into some serious wet weather, most brands up to the 27-gallon range will fit inside of a contractor-style garbage bag. It never hurts to keep a few extras in a tub, whether for wrapping around the outside or for placing the interior contents inside the bags to keep them dry. Stackable, or nesting, tops are most efficient for consistency and maximizing storage footprint. Heavier tubs won't slide or move around when they have elevated lids or lips and are stacked on top of each other. For those nervous about prying eyes (or high winds) when parked to refuel or when that breakfast burrito hits, consider tubs with handle slots for locks.

These tubs make ideal camp kits as well. They keep food dry and out of the sun, and the larger versions can easily fit bigger cutting boards and cooking tools. The same theory applies to airbeds, sleeping bags, pillows, and compressible items that you want to keep easily stored away. A loose sleeping bag rattling around a vehicle only takes up space and may result in some rips or spills on the bag. Keep it packed away in a tub, safe and clean. These are equally beneficial for those looking to simply congregate tents, shells, tarps, poles, and other tent items together. Tent stakes and spikes can easily slip through milk crates and puncture tent walls when packed together. With the ease of labeling these rubber tubs, you can easily differentiate between warm-weather and cold-weather tents.

The trick here is buying the same brand to keep them stackable and the organization systems continuous. For smaller miscellaneous items, the small

This fly box has seen a lot of time on the road during the season, which is evident in the disorganized flies. While flies in random order in an individual box can be a minor annoyance, this small box in the back of a truck or tackle bag can be more problematic if it gets lost or left behind. PHOTO BY OLIVER SUTRO

clear boxes can be ideal. While anglers should always consider reducing overall tackle and packed items, for those who find safety in being overly prepared, these clear tubs are ideal. With the ability to nest inside of a larger tote, they are great for extra on-the-water tools, boot laces, sunscreen/bug spray, and anything else that's small and likely to get jostled around and lost during transport. For those who find themselves attaching and removing wading aides, or cleats, this is a suitable place to pack them away so you won't lose spikes or other sharp items inside of your normal travel bag.

Like a safety blanket, these smaller containers will serve to make finding smaller and backup items easy, while also giving you the peace of mind to be able to recheck everything before hitting the road.

PACKING CUBES

Packing cubes—soft-sided, zippered, and often made from a water-resistant type of polyester—can be a powerful addition to the packing and repacking scenario. Most modern packing cubes, which come in a clamshell-type design with a netted, zippered divider (or two) on the inside, do allow for very efficient and nondescript packing of fly boxes, leaders, tools, and other items that can easily hide in the corner of a duffle bag.

With the addition of a few ziplock bags, they can keep away moisture and keep prying eyes from easily stolen items. More important is just keeping the loose items from slipping into cracks and crevices, never to be found during unpacking.

The collapsible nature of packing cubes can be helpful for clothing items as well, to keep your socks, buffs, underwear, and toiletry items neat and packed away. They undoubtedly have aided in more compact, efficient packing. They can be purposed in a lot of ways and even make good gear holders in backpacks or boat bags during a float.

Pack and Ship

While we've all faced the uncertainty of packages arriving, and in what type of condition, packing and shipping tackle and accessories is something that anglers should always consider. Of course, location, time of delivery, and time spent fishing in the area can be limiting factors, but shipping larger, bulkier items like waders, boots, fly reels, and fly boxes can make on-the-ground travel much easier. This allows for smooth and efficient footsteps and maneuvering around busy airports and in crowded rental cars, and it also means less to carry on and leave to the whims of recalcitrant baggage handlers.

However, this method does have the potential to incur some difficulties. While we mean no disrespect to the blue-felt short-clad foot soldiers of the US Postal Service, shipping through the current USPS is the least palatable option for anglers in this situation. Major carriers like FedEx and UPS can provide real-time tracking data and typically a more robust network of delivery options in the often truncated timeline that an angler may have in a location.

What shipping items beforehand can do outside of ease of movement and less physical weight during the travel process is allow anglers to transport securely and safely expensive tackle from location to location with a limited amount of potential failure. When we say failure, we mean the inherent and uncompensated damage that occurs when others handle or unpack your gear during airport examinations.

We've been there many times. We thought we cleverly packed away an expensive reel or rod, deep down in a duffle bag, only to open it up on arrival to notice the frame is dented or the rod has been jammed haphazardly into the rod sock. Even worse, that pair of pliers you hid away is just gone. No recourse, no compensation, and it usually leads only to endless phone calls and hollow threats about abandoning support of the airline and taking your business elsewhere. We can assure you that the customer service technician sitting in a long row of cubicles at the airport in Philadelphia doesn't care how your 1950s Hardy Princess spool got bent or what happened to your new Hatch pliers while flying to Boise, Idaho.

It's only going to be an exercise in futility, and you'll still be stuck sitting on a motel bed cussing out someone two thousand miles away, while not fishing.

Spare Rods and Parts

Whether human error (which is 98 percent of the time) or some perceived slight from the fly-fishing gods, rods do break, guides fail, and taking only one rod on a trip can cause intense anger. A lot can go wrong in fishing scenarios, and simply the best course of action is to not travel with just one rod. Whether it's a careless baggage handler, an errant car door or window, or just what happens after years of hard use on a rod, sections can break. It happens.

What's worse is either being stuck on the sidelines without a functioning rod or scrambling to try to buy a replacement in town.

For the most part, unless on an extended trip, getting a rod replaced or a new section mailed to that tiny mountain town would take all but a miracle. It's a stress-laden proposition and one even the most focused customer service agent would struggle to make happen in time. So how can you avoid this disaster while on your trip out West? The simple answer is to always travel with multiple rods. Having a backup, or backup to the backup, allows you to continue your fishing trip. As noted earlier, a variety of rod options will also add to the diversity of the toolbox. They can likewise provide the bridge for when a rod is forced from the roster.

While it is strongly suggested that anglers prepare different rods for different techniques, we are also cognizant of budget constraints. The price tags for multiple rods and reels can add up, and simply adding tackle is not a linear path to success. However, as manufacturing continues to improve at impressive rates, many companies are now offering replacement sections for sale. The ability to purchase spare sections can quickly salvage a trip. While rods that come with an extra tip section may seem like a gimmick, there are benefits to having this extra piece when failure inevitably strikes.

The warranty and the parts and service abilities of rod companies should always be examined when evaluating your next purchase. Spend some time looking around, paying attention to replacement delivery timelines, individual fees, and which companies offer additional sections for sale.

Whether a single, serious investment and a tool you plan to spend many hours with or a backup that may end up in the closet with a dozen other rod tubes, rods are an investment. While space-age materials, fancy guides, and colorful blanks are cool, anglers can do themselves long-term favors by digging deeper into the different brands and their replacement or warranty policies.

Protecting the Rod on the Road

It is critical to consider how to handle your rod while on a trip to ensure it gets to the destination in working order. While the single rod tube that likely came with your rod will suffice, the ability to critically select multi-rod tubes, rod/reel carriers, and the many other traveling accessories now offered by manufacturers can make the experience all the better.

Like a grizzled general, diverting forces and hitting a target from multiple angles, anglers should consider the same strategy of dividing and conquering when traveling with tackle. Splitting up vital tackle across bags can keep things safe and save you if a bag doesn't make it or someone goes rifling through it. Flies, leader tippets, and even rods and reels can be replaced while on the road. Boots, waders, and some of those more sentimental items aren't always as easy to do without.

Depending on the type of trip, the number of days of travel, and the solitude of the location, I (Nicholas) try to separate my rods into at least two points of contact. For 75 percent of my air travel, I carry a set of rods and typically my waders and boots. Partially due to the fact that I have long legs and large feet (with a relatively average waist and set of shoulders), finding a replacement set of boots and waders, even in densely populated areas, is a challenge. Tackle should be viewed the same as medication: Keep close what cannot easily be replaced. Don't trust the airlines and baggage handlers.

In a rod roll-up or multi-rod tube carrier, I will pack as many rods as safely feasible. If I can hold it, I feel safe. Probably some deep-seated issue from my childhood, but I digress. With these items, and usually a reel or two packed away inside of my wading boots, I know I will be able to get through the trip should disaster strike and my large bags are delayed.

Traveling with multiple rods ensures that you always have a backup. Even if conditions or techniques don't change, having a spare rod makes for one less trip to the truck. PHOTO BY OLIVER SUTRO

Despite the constant issues, or more likely blatant confusion and lack of concrete information from major airlines, the size and type of rods you can carry on a plane is often cloudy. After over twelve years of flying both domestically and internationally with way too many rods of varying lengths, the best advice I can offer is to go for it. I've never once been challenged with a rod case or long tube (especially traveling internationally with double-handed rods over 15 feet in length). Usually, there is space somewhere to store it. With a smile and a simple ask, it usually works out. If anything, it tends to be a talking point and you've made a nice, albeit brief, connection with a fellow supporter of the outdoors. Of course, people have bad days, and I can't promise every bizarre length and configuration of a fly rod tube is going to find a welcome home on a plane. But odds are if you drag along a standard-length rod, there will be a safe spot for it.

Worst-case scenario, you spend a few extra bucks and check it. You've come this far; is another $50 going to stop you?

My standard for checked items is a long, nearly 58-inch, polymer rifle-style case. With wheels and sufficient length for rods up to 13½ feet, I can pack a healthy amount of rods safely and securely. While my case has many years of stickers blotted across it like the skin of a punk rock drummer, it still gets looks and questions regarding its contents. When I inform people, "No, it's not rifles or guns; it's fishing rods," they tend to absently stare to the side as if I asked them whether they had a minute to discuss the environment. Fishing rods just don't solicit the reactions that an AR-15 does these days.

With lockable latches, I can ensure the contents are (mostly) safe, and it goes through special screening measures. It gets better treatment than the typical checked bag.

Several soft- and semihard-sided cases and travel bags dedicated to fishing tackle are available. Many leading brands offer rectangle-style cases that are perfect for trips that require only a handful of rods and reels. With dimensions ideal for a carry-on bag, you can safely keep an eye on the tackle. The four- to six-rod carriers come in a variety of constructions and sizes, with more cases expanding in length to accommodate rods up to 9½ feet.

What to Do When the Party's Over

While it may seem easiest just to shove all that wet gear, kinked-up leaders, and loose flies away when breaking down camp, fight the urge and take a few minutes to put up your gear. Not only will doing so help to better preserve and care for items such as boots, waders, and jackets, but it will also make the unpacking and reorganizing much more efficient.

The post-trip wrap-up here can be just as beneficial as the committed time you spent before the trip practicing your casts and other essential work.

Even in an ideal sunny, dry environment, it is advised to hang-dry important items like waders and jackets. Despite spending over $700 on that fancy pair of multi-millimeter, abrasion-resistant Gore-Tex waders with anatomically engineered neoprene stocking feet and anti-microbial coatings, waders still can be a sensitive item. They require care and attention. For the occasional angler, that type of care can seem a bit over the top. Ignore the ease of quickly folding those wet and sandy waders and shoving them into a dark bag.

SIMPLE MAINTENANCE

By simply hanging your waders, hosing them down with fresh water, and letting them sit in the air to dry, you'll likely add trips and even years to their usability. The same is true for wading jackets. This is a good time to dig into each opening and remove the tangled leaders, bunched-up flies, and candy bar wrappers from the pockets.

One of the benefits (but also downfalls) of modern technology is the addition of pockets. Designed for ease of use, they often are overused or over-installed onto an item and create more hiding spots for tackle. Empty the pockets, pull out what is still useful for the next trip, and recycle the rest.

This is also a good time to perform some simple zipper maintenance. While various manufacturers either extol or condemn the use of waxes or other lubricating agents on zippers, anglers should at least make the effort to clean out gritty debris and wash well-worn zippers. Doing so now, in an outside environment, will prevent sand, dirt, and mud from securing itself within the tongues and grooves. Compressed air canisters, like those commonly used on computer keyboards, can also be helpful for getting into hard-to-reach corners and removing potentially zipper-jamming particles.

When doing a post-trip washing down of wading boots, there are a few areas to focus on. Remove the detritus from the soles and clean off any grit or caked-on mud from spikes, bars, and other attached wading aides. Even after just a few hours of sitting in a bag and drying on a plane ride, these items can become adhered to the sole.

Proactive maintenance of waders, by keeping them hung in an appropriate environment, will undoubtedly add years to their utility. An inappropriate place to store waders is folded or crammed into a bag for long periods of time. Storing waders outside or in other potentially volatile environments can lead to fading and cracking.
PHOTO BY NICHOLAS CONKLIN

Take some time now and save a dusty, dirty hassle on your return home. And if you're not lucky enough to have a separate fishing room or garage to handle this dirty work, your live-in partner will thank you for not dragging this shit through the house and washing it in the kitchen sink.

The second stage is the removal of bootlaces. While seemingly time-consuming, and rarely done by professionals, this can greatly extend their life span. Often overlooked by venturing anglers, laces, especially of this class and size, can be hard to find. You don't want to be stuck wandering the aisles of a small-town grocery store looking for replacement laces, only to find white laces meant for some old guy's New Balance sneakers. Take care of your laces, and travel with a backup pair (or two).

With all the focus on the outside of the boot, one simple hack can help prolong the inside of the boot as well. During the many trips to town for beer and ice, stop and grab some newspapers. Yes, newspapers—antiquated, printed sheets that are used to report actual news and shine a light on the abuses of government officials and those in power. Nowadays, they often sit idly in front of grocery stores or on the side of the street in metal tombs and cater to news of festivals and real estate opportunities.

However, a section or two of the local rag can be great for attracting moisture. Balling up the obituary section and sticking it into the damp recesses of your wading boots will help pull moisture out and keep your boots dry and fresh-smelling. Other moisture-destroying items like desiccant packets can be great for keeping overly damp gear manageable. For those fishing in wetter or snowy conditions, these can work double time in tackle bags, packs, and fly boxes as well as in boots and waders.

If storing boots during a trip, or after an especially long day submerged, try to hang or orient your boots upside down. This will prevent water from puddling inside and allow for a little more comfort when slipping boots on during those cold, frosty morning fishing sessions.

This is also a good time to apply some simple line cleaning. Like all post-trip cleaning, odds are you let your lines and leaders sit on the spools for days, sometimes weeks or, shame-shame, seasons until you address the maintenance. As covered in previous chapters, the quality of your fly line is of enormous

Wet boots and well-used gear should always be cleaned and dried for a safe return trip home. Just because you set them outside the night before doesn't mean they are dry. Suffer through the smells and dampness and make sure they are dried indoors or somewhere near a heat source that could help speed the process along. PHOTO BY NICHOLAS CONKLIN

importance to your entire system. So why then is it one of the more neglected pieces of equipment?

For a quick, post-trip, prepack cleaning, simply run your fly lines through some restaurant-style moist towelettes. Not just for cleaning your greasy, meat-stained fingers after lunch, this is an inexpensive and effortless way to wipe surface grime and debris from a fly line. They can be purchased in bulk or swiped from your local BBQ joint, they pack easily, and again they are cheap (or free), so there shouldn't be much hesitation about their single use. We do typically suggest staying away from wipes or cleaners that tout "anti-virus" or antibacterial components. While our research and testing have not found any deterioration of a fly line, that could happen, and the harshness of the smell isn't one you want sticking around the reel.

While we have not uncovered any empirical data to suggest that fly lines can be carriers of waterborne invaders like waders and boots can, it makes sense that something could attach itself to your line and you thus become an unwilling transporter of some waterborne disease. If done right, your fly line will spend almost as much time in the water as your boots, waders, and/or boat, so why not be cautious and make every attempt to clean your line when hopping between rivers and lakes?

Of course, fly line–specific cleaners are widely available. Many come with high-quality pads that can further pull any grit from the line, and the cleaning agent also has a slicking component. As always, we suggest a good, standard dish soap and warm water cleaning after the trip to ensure as much of a lifespan to a line as possible.

FLY BOX CLEANING

While not an immediate post-trip requirement, taking some serious time in between trips or seasons to reorganize and replace flies is key for the angler looking to improve their game. Organizing and replacing flies, noting which flies were hot and which were not, can shape the overall footprint of your boxes.

This downtime is also a good chance to go through and realign those wayward flies that may have been put in different slots or in different boxes altogether during the height of a busy fishing day. This post-trip fly box time is also a chance to refine your efficiency in locating, selecting, and tying on a desired pattern.

A study of post-trip fly boxes has led to the notion that most anglers fail to examine and reexamine how and where they place their flies. Most just take that wad of flies from the plastic hockey puck and dump them wherever, which ultimately leads to a variety of patterns, colors, and misplaced flies. The same scenario arises throughout a trip; flies move around, and it is easy to misplace or underestimate the number of flies you have. That light purple midge, in size 20, may be hiding under another, larger pattern.

The angler should orient flies by how they will most comfortably hold the box. By being overly considerate of location, based on most used patterns and sizes, you can further subtract searching time and fumbling of boxes. If right-hand dominant and you can comfortably and safely manage the box in your left hand while selecting the desired fly with

Anglers should always keep in mind when replacing or building out fly boxes that when it's time for actual application, they will be on the water or sitting along the bank. While it may seem easy while sitting in the comfy confines of your home, fly orientation and replacement can be critical to efficient movements while retying in actual fishing scenarios. PHOTO BY OLIVER SUTRO

Remember that fly fishing is supposed to be fun and worth protecting. Most of us started angling simply, with big smiles and struggling to grab slippery fish. We need to remember that childlike wonder. Awareness of the impact on the resource should be taught early and thought of often. PHOTO BY NICHOLAS CONKLIN

your right hand, you will tend to load the slots on the right side. For hinged or multilayered fly boxes, which can require a bit more maneuvering, you may want to keep your most used patterns on the first slat, and then to the left, with your less used, more specific flies on the opposite side of the leaf.

While misplacing or simply overlooking that effective fly that somehow went missing can be an on-the-water issue, it can also distract and add to the trip-to-the-fly-shop time. Extra time and money seem to compound when anglers are unable to locate the fly that seems to be working that day. This is another convincing argument for limiting the overall number of patterns by focusing on a handful of simple flies in varied sizes. Don't let the massive guide boxes in fly-fishing manufacturer images fool you—effectiveness and increased fun can be had with a streamlined approach. Little flies can be deceptive, so opportunistic anglers should always be looking for ways to refine and reorganize their boxes.

When organizing fly boxes following a trip, this is also a good time to dig around in all the pockets and corners for any loose fly pucks. Regardless of how careful you think you were, you'll likely have stashed something in a wader flap, side pocket, or elsewhere. You spent the time and money to purchase these flies at the shop, so there must have been some compelling reason to get them. Don't let them die a slow death in a lost plastic box or at the bottom of a boat bag.

INDEX

P

R

ABOUT THE AUTHORS

Nicholas Conklin

Nicholas Conklin is the fly-fishing product designer for the Texas-based company Temple Fork Outfitters. He is the chair of the Board of Directors for the American Fly Fishing Trade Association (AFFTA) and previously served as treasurer and chair of the Show and Summit committee. He has also served on several committees focused on the sustainable growth of fly-fishing-related businesses.

Nick is also an accomplished tournament two-handed rod caster, having competed at Spey O' Rama (San Francisco, California) and the Celtic Spey Challenge (Belfast, Ireland). He has presented casting demonstrations at many national shows such as the Fly Fishing Show, the Midwest Fly Fishing Expo, Spey Nation, and the Sandy River Spey Clave.

Nick's articles and photography have appeared in *Field & Stream*, *The Drake*, *Lone Star Outdoor News*, the *Fort Worth Star-Telegram*, and other regional and national publications.

Originally from Michigan, Nick grew up fishing for trout and bass in and around the Great Lakes. After spending much of his youth fishing iconic waters like the Pere Marquette, Au Sable, and Grand Rivers, he moved to Texas to chase salty Gulf Coast fish. He currently resides in Dallas but spends much of his time on the coast or roaming around the western United States.

Capt. Collins D. Illich

Capt. Collins D. Illich manages operations and design for the Texas-based company Sawgrass Fishing Rods. Additionally, he operates a light-tackle sport-fishing operation in North Texas.

Collins holds a 25T Master Captain's License from the United States Coast Guard and is an IGFA-certified Captain. He has hosted anglers on remote exploratory trips worldwide, exploring fisheries on six continents, and is enthusiastic about finding new, undiscovered water. If he is not looking for a new fishery, he is likely fishing southwest Montana with a high likelihood he is on the Upper Madison.

Originally from Texas, Collins was raised fishing the Gulf Coast and Northern Rockies. He currently lives in Dallas with his wife, three kiddos, and his loyal Labrador.